The Quality Calibration Handbook

Also Available from ASQ Quality Press:

The Metrology Handbook
Jay L. Bucher, editor

Make Your Destructive, Dynamic, and Attribute Measurement System Work for You
William D. Mawby

ANSI/ISO/IEC 17025-2005: General requirements for the competence of testing and calibration laboratories
ANSI/ISO/IEC

ANSI/ISO/ASQ Q10012-2003: Measurement management systems—Requirements for measurement processes and measuring equipment
ANSI/ISO/ASQ

The Uncertainty of Measurements: Physical and Chemical Metrology: Impact and Analysis
S. K. Kimothi

Managing the Metrology System, Third Edition
C. Robert Pennella

Get It Right: A Guide to Strategic Quality Systems
Ken Imler

Integrating Inspection Management into Your Quality Improvement System
William D. Mawby

Failure Mode and Effect Analysis: FMEA from Theory to Execution, Second Edition
D. H. Stamatis

Root Cause Analysis: Simplified Tools and Techniques, Second Edition
Bjørn Andersen and Tom Fagerhaug

The Certified Manager of Quality/Organizational Excellence Handbook: Third Edition
Russell T. Westcott, editor

Leadership for Results: Removing Barriers to Success for People, Projects, and Processes
Tom Barker

To request a complimentary catalog of ASQ Quality Press publications,
call 800-248-1946 or visit our Web site at http://qualitypress.asq.org.

The Quality Calibration Handbook

Developing and Managing a Calibration Program

Jay L. Bucher

ASQ Quality Press
Milwaukee, Wisconsin

American Society for Quality, Quality Press, Milwaukee 53203
© 2007 American Society for Quality
All rights reserved. Published 2006
Printed in the United States of America

17 16   5 4

Library of Congress Cataloging-in-Publication Data
Bucher, Jay L., 1949–
  The quality calibration handbook : developing and managing a calibration program / Jay L. Bucher. – 1st ed.
      p. cm.
  Includes bibliographical references and index.
  ISBN 0-87389-704-8 (casebound : alk. paper)
  1. Mensuration—Handbooks, manuals, etc. 2. Calibration—Handbooks, manuals, etc.
3. Quality assurance—Handbooks, manuals, etc. I. Title.

  T50.B75 2006
  658.4'013-dc22   2006027515

  ISBN–10: 0–87389–704–8
  ISBN–13: 978–0–87389–704–4

Publisher: William A. Tony
Acquisitions Editor: Matt Meinholz
Project Editor: Paul O'Mara
Production Administrator: Randall Benson

ASQ Mission: The American Society for Quality advances individual, organizational, and community excellence worldwide through learning, quality improvement, and knowledge exchange.

Attention Bookstores, Wholesalers, Schools, and Corporations: ASQ Quality Press books, videotapes, audiotapes, and software are available at quantity discounts with bulk purchases for business, educational, or instructional use. For information, please contact ASQ Quality Press at 800-248-1946, or write to ASQ Quality Press, P.O. Box 3005, Milwaukee, WI 53201-3005.

To place orders or to request a free copy of the ASQ Quality Press Publications Catalog, including ASQ membership information, call 800-248-1946. Visit our Web site at www.asq.org or http://qualitypress.asq.org.

∞ Printed on acid-free paper

Quality Press
600 N. Plankinton Avenue
Milwaukee, Wisconsin 53203
Call toll free 800-248-1946
Fax 414-272-1734
www.asq.org
http://qualitypress.asq.org
http://standardsgroup.asq.org
E-mail: authors@asq.org

ASQ
AMERICAN SOCIETY
FOR QUALITY

Contents

Figures and Tables

Preface

M an initially invented the wheel long before records were officially kept or written down. Yet it is very possible that variations on the wheel were reinvented, modified, changed, and improved upon many times, in many parts of the world, over many generations. If records had been kept for all to access, and they were available when needed, less invention might have taken place, and more "improvement" could have taken place.

The same can be said for calibration. The good folks in the metrology and calibration fields have been inventing ways to calibrate (calibration systems vice calibration of a particular item) for as long as I can remember, and that goes back to 1971 when I started my career in metrology and calibration. Each branch of the military has its way of performing calibrations. Each third-party calibration lab has its way, along with each department or calibration function within most companies throughout the world. When all is said and done, is there one formal way to calibrate? Hopefully not, because there are many factors that go into putting together a calibration system, most depending on what standards or regulations govern the calibration function. However, is there one tried-and-true quality calibration system that every organization can use as a foundation for its personalized program? Yes, there is. That is where *The Quality Calibration Handbook* comes in.

By using the quality calibration system outlined and demonstrated here, any organization can put together its own version to meet its specific requirements and/or regulations. Organizations can avoid having to completely reinvent the critical wheel called calibration. But we're getting ahead of ourselves. First, we have to define what calibration is and what it is not.

In 1998, during an initial audit for ISO 9001, a trained, experienced auditor was inspecting a metrology department. While reviewing some of the calibration records, the auditor continued to refer to calibration as the adjustment of the test equipment. She said that the liquid-in-glass (LIG) thermometer could not be calibrated because it could not be adjusted. Her audit contained three instances of this type of statement. I tried to remain as patient and sympathetic of her ignorance as humanly possible (there have been times in the past when I have not been known for extreme patience). After the referral about the LIG thermometer, there was no choice but to educate the auditor about the real meaning of calibration. The department manager, company management representative, and others in attendance all took a deep breath and held it. They thought there was never a time that one corrected an auditor. Under most circumstances, that might be true. But I could not allow her to be totally incorrect in how she did her job in this circumstance. I explained to the auditor—using the VIM and NCSLI definitions and examples from those definitions—what calibration really meant. She said she would make a note of the true definition so other auditors could be educated, and she was grateful for the education. In that case, all's well that ends well. There were no findings, observations, or write-ups. More information can be found in Chapter 20 on how to work with auditors. There is also advice for what to say or not say during audits.

Eight years later, in April 2006, during a meeting with the regional sales representative of a test equipment company, a very similar incident happened. We had owned their products for several years, calibrated them on a yearly basis, and decided to upgrade to a new version of the product. A sit-down demonstration was arranged. The first statement from the salesman was, "The new version of the software allows calibration of these items." I apologized and said that we have been calibrating our older models for years. He said that was not possible because the previous versions of software did not allow for any type of adjustment. I explained what calibration really meant, and he looked at me like I had grown horns. It would appear the word still has not gotten out about the true meaning of calibration. So, without further ado, here it is.

The following primary and secondary definitions are according to VIM 6.11 [an acronym commonly used to identify the ISO *International Vocabulary of Basic and General Terms in Metrology (VIM)*, also known by the French title, *Vocabulaire international des termes foundamentaux et généraux de métrologie (VIM)*] and NCSL International (pp. 4–5):[1]

> *Calibration* is a term that has many different-but-similar definitions. It is the process of verifying the capability and performance of an item of measuring and test equipment by comparison to traceable measurement standards. Calibration is performed with the item being calibrated in its normal operating configuration—as the normal operator would use it. The calibration process uses traceable external stimuli, measurement standards, or artifacts as needed to verify the performance. Calibration provides assurance that the instrument is capable of making measurements to its performance specification when it is correctly used.

> The result of a calibration is a determination of the performance quality of the instrument with respect to the desired specifications. This may be in the form of a pass/fail decision, determining or assigning one or more values, or the determination of one or more corrections.

> The calibration *process* consists of comparing test equipment with specified tolerances but of unverified accuracy, to a measurement system or device of specified capability and known uncertainty, in order to detect, report, or minimize by adjustment any deviations from the tolerance limits or any other variation in the accuracy of the instrument being compared. Calibration is performed according to a specified documented calibration procedure, under a set of specified and controlled measurement conditions, and with a specified and controlled measurement system.

Many manufacturers [auditors, QA inspectors, and so on] INCORRECTLY use the term *calibration* to name the process of alignment or adjustment of an item that is either *newly manufactured* or is *known to be out of tolerance*, or is otherwise in an indeterminate state. Many *calibration* procedures in manufacturers' manuals are actually factory alignment procedures that need to be performed only if a UUC (unit under calibration) is in an indeterminate state because it is *being manufactured, is known to be out of tolerance,* or *after it is repaired. When used this way, calibration means the same as alignment or adjustment, which are repair activities and excluded from the metrological definition of calibration.*

Here is the bottom line when it comes to what calibration is: A *comparison* of test equipment with an *unknown uncertainty* to a standard with a *known uncertainty. Calibration* is the comparison of a piece of test equipment with a standard, regardless of whether the standard is kept at NIST. It is the reference standard used by a third-party calibration lab or the working standard used every day by calibration technicians. *It is a comparison.* You need something to calibrate and a standard to compare it against.

Calibration has nothing to do with adjustment, repair, alignment, zeroing, or standardizing. All of these can be incorporated into the process at some point, depending on what the item is, how it is used, and in some cases, at what level it is being calibrated.

Here are some quick thoughts about alternative titles for this book:

Calibration—How to Make Friends and Scare the Hell Out of Your Competition
Calibration—No Fat, No Sugar, No Calories . . . Nothing but Profit
Calibration—It's No Longer a Dirty Word
Calibration or Bust!

If it seems that I am excited about this book, it's because I am. I *am* excited about calibration and metrology, and how they impact our quality of life. I want to shout it from the mountaintops, splash it across newspaper pages, and discuss nothing else with family and friends. Of course, that would get boring to the uninitiated, or to everyone who does not understand how critical calibration is to everything around us. So, I have to temper my enthusiasm and write a book about it.

Part of the reason for writing a book about calibration is to educate the general public about the great need for a quality calibration system. One of the distinguished reviewers of this book's original abstract was so bold as to pontificate: "Simply stated, the title is boring. But in all fairness, there isn't much pizzazz to be found in discussing calibration systems." Nothing could be further from the truth.

Here is what a few calibration technicians might have to say about the impact their jobs have on our overall quality of life:

- The equipment I calibrate on a daily basis helps catch killers and rapists all across America.
- My work was instrumental in helping to set the innocent free from prison.
- Airline accidents and mishaps are down in direct proportion to the accuracy of my work.
- The highways and byways are safer because of the due diligence of our calibration program in automotive manufacturing.

- The number of people helped by the medications we manufacture using calibrated systems and measurements is incalculable.
- There are more new discoveries in drugs and cures for the incurable than ever before, in part thanks to the repeatable readings and accuracy of our calibrated test equipment.

How is all of this possible? Quality calibration systems are the foundation for improving research and development (R&D), production, and quality assurance arenas through accurate, reliable, and traceable calibrations of their test equipment. If quality calibration were not important, then it would not be a requirement in industry, government, and the private sector.

I help catch killers and rapists, all the while aiding in setting the innocent free. I do this every day. My cohorts, other calibration practitioners, do similar life-saving work to prevent air disasters, vehicular crashes, and poisonings. How is this possible? By my ensuring the calibration of test equipment used in the production of genetic identity kits used by law enforcement at crime scenes, the guilty are often caught and the innocent exonerated. Calibrated test equipment used in support of the airline and automotive industries helps prevent disasters. While calibration technicians do their seemingly boring, mundane jobs at the nation's pharmaceutical companies, they are quietly laying the foundation for quality treatments that keep all of us healthy and help cure diseases and sometimes prevent death.

Not much pizzazz in any of that? It's time someone woke up and smelled the coffee. This book explains why a quality calibration system can be the difference between life and death, success and failure, and—most important to shareholders and boards of directors—profit and loss.

NOTE

1. Jay L. Bucher, *The Metrology Handbook* (Milwaukee: ASQ Quality Press, 2004), 472.

Acknowledgments

I would like to express my thanks to my father and mother. Even though they both passed away several years ago, their influence and work ethics have kept me in good stead these many years. Most of us don't appreciate our parents until after we lose them, sometimes not even then. I was fortunate to have known ahead of time that they were special. My dad only had a ninth-grade education, but through hard work and a continuing desire to learn, he became very successful. He shared his knowledge, wealth, and wisdom with family and friends. I'm hoping to emulate him in a small way by writing this book. Thanks, Gramps, for the great example to follow.

I'd also like to thank Lou Mezei, senior scientific fellow and computer programmer extraordinaire, for his friendship, inspiration, and drive to excel at any task. I've said it many times, in many ways, but the bottom line is: Lou, "you da man." Lou has the ability to take any problem, task, or project and produce an end result second to none. Not only is the final result better than what was requested, but future requirements and objects are automatically added to preclude having to reinvent the wheel somewhere down the line. It is indeed a privilege to observe a true genius in action. Thanks, Lou, for allowing this old Minnesota hog farmer to watch you in action.

Thanks also go out to William A. (Bill) Linton, president/CEO of Promega Corporation, for giving me the opportunity to work and play at his company. It is a privilege to work in an organization that stays on the cutting edge of innovation and technology. It keeps the rest of us on our toes, motivating us to keep discovering new and better ways to do our jobs. I suppose the best I can say in just a few words is: It's a great place to work. At least it is for Team Metrology and me.

Speaking of Team Metrology, thank you, Keela and Karl. Your help and support through the years and with this book have given me a brighter light, not only to see where I'm going, but more importantly, to see the best way to get there. Together we have jumped many hurdles, found new and better ways to get the job accomplished, and still found time for the occasional laugh and ice cream. You bring new meaning to our motto of providing quality service in a timely manner. Without either one of you, it would be difficult to say, "We don't have problems . . . we have solutions!" I feel privileged to call both of you friends.

And thanks to the Fischbecks. Bob "Oyabun," for giving my wife the opportunity of a lifetime. It brought us to "'Sconsin" and helped to open so many doors we had to move to a bigger house. Thanks also to Pat, the greatest innkeeper in the world, for your friendship and for expanding my exposure to authors of historical romance novels.

Last, but not least, I must thank my family for their patience, understanding, and support. They stood by me through thick and thin, and never let me doubt what I was doing or why I was doing it. Both of you keep me young at heart and constantly on my toes. To have the love and respect of my daughter and wife is something I have to work at every single day. I don't take that for granted, nor should I. As the lyrics from Spiral Starecase say, "I love you more today than yesterday, but not as much as tomorrow."

To all the calibration technicians, supervisors, and managers in the Army, Navy, Air Force, Marines, and Coast Guard, thank you for doing what you do to help keep our great country second to none. I know from personal experience that you seldom receive thanks or appreciation for doing the mundane and repetitive tasks that eventually result in somebody else getting credit for a job well done. Accuracy and precision still have their place and all of you, in some measure, ensure the job is done right the first time. "Close enough for government work" is still not part of your mantra—thank God! I hope this book helps a few of you realize that if you have only served for a few years or an entire career, there is life after PMEL. Godspeed to all our brothers and sisters serving in the military.

Jay L. Bucher, MSgt, USAF (Ret)
ASQ Sr. Member, CCT

Part I
Why Calibration Is Critical

1
Preventing the Next
Great Train Wreck

On January 21, 2005, during the annual Measurement Science Conference (MSC), held at the Disneyland Convention Center in Anaheim, California, the Measurement Quality Division (MQD) sponsored a seminar on metrology education. During that seminar, one of the audience participants was Dr. Klaus-Dieter Sommer of Germany.

Dr. Sommer explained that he was a guest professor at a university in China. He told how the university had an input of 8,000 students every year, and they were all studying measurement techniques within a metrology system. He said the Chinese government has increased their attendance to around 12,000 students in metrology and measurement techniques. His question then was, "Why doesn't the United States and/or Germany train and educate in the field of metrology and calibration they way the Chinese are doing?"

After one person gave their ideas, I had an epiphany. I raised my hand and answered, "Because we haven't had the great train wreck yet." We haven't had a train wreck where someone says, "Calibration was the problem." We haven't had the train wreck where there is a great loss of life or limb, or many businesses go bankrupt, or a great many people lose their livelihood or retirement funds.

The automotive industry has had many train wrecks. We all remember the problems that Ford Motor Company and Firestone had a few years ago. The airline industry has had many instances of tragedy and loss of life over the years. The nuclear industry has had its fair share of problems, too.

Without calibration, or by using incorrect calibrations, all of us pay more at the gas station, for food weighed incorrectly at the checkout counter, and for speeding tickets. Incorrect amounts of ingredients in your prescription and over-the-counter (OTC) drugs can cost more, or even cause illness or death. Because of poor or incorrect calibration, killers and rapists are either not convicted or are released on bad evidence. Crime labs cannot identify the remains of victims or wrongly identify victims in the case of mass graves. Airliners fly into mountaintops and off the ends of runways because they don't know their altitude and/or speed. Babies are not correctly weighed at birth. The amount of drugs confiscated in a raid determines whether the offense is a misdemeanor or a felony; which weight is correct?

Errors in calibration can effect the automotive, nuclear, or space industries. They can also have an impact on how long or wide a 2 × 4 is, not to mention the thickness of drywall, how much radiation is emitted by a microwave oven, how much money you are overcharged each month because your gas meter is turning at the wrong speed, or how much extra a company pays for their stamps because the scale at the post office adds 5 percent to package weight. No more watching your favorite television shows or listening to your favorite songs or talk radio, because due to calibration errors, the frequency would be off enough that TVs and radios would be useless. Satellites and everything they affect would be a thing of the past, as would be the manufacturing and production of almost everything made in the world today.

The United States Food and Drug Administration (FDA), an agency that protects the health of the American people, is one of the most successful and proudest creations of the American democracy. The FDA was created in the early 20th century amid revelations about filth in the Chicago stockyards that shocked the nation into the awareness that in an industrial economy, protection against unsafe products is beyond any individual's means. The U.S. Congress responded to Upton Sinclair's best-selling *The Jungle* by passing the Food and Drugs Act of 1906, which prohibited interstate commerce in misbranded and adulterated food and drugs. Enforcement of the law was entrusted to the U.S. Department of Agriculture's Bureau of Chemistry, which later became the FDA.

The act was the first of more than 200 laws that constitute one of the world's most comprehensive and effective networks of public health and consumer protections. Here are a few of the congressional milestones:

- The Federal Food, Drug, and Cosmetic (FD&C) Act of 1938 was passed following the death of 107 people, mostly children, who took a legally marketed poisonous elixir of sulfanilamide. The FD&C Act completely overhauled the public health system. Among other provisions, the law authorized the FDA to demand evidence of safety for new drugs, issue standards for food, and conduct factory inspections.
- The Kefauver-Harris Amendments of 1962, spurred by the thalidomide tragedy in Europe (and the FDA's vigilance that prevented the drug from being marketed in the United States), strengthened the rules for drug safety and required manufacturers to prove their drugs' effectiveness.
- The Medical Device Amendments of 1976 followed a U.S. Senate finding that faulty medical devices had caused 10,000 injuries, including 731 deaths. The law applied safety and effectiveness safeguards to new devices.

Today, the FDA regulates an estimated $1 trillion worth of products a year. It ensures the safety of all food except for meat, poultry, and some egg products; ensures the safety and effectiveness of all drugs, biological products (including blood, vaccines, and tissues for transplantation), medical devices, and animal drugs and feed; and makes sure that cosmetics and medical and consumer products that emit radiation do no harm.[1]

But there has not been a calibration train wreck . . . yet. Until there is, people will not realize that they need a quality calibration system.

Today, it is even more important that we be proactive in our application of quality calibration programs. If industry waits for the great train wreck, it will be too late. The pendulum will swing too far and we will have to abide far more government control than is needed. Recent history provides myriad examples of the need.

Hurricanes were killing hundreds of people along the gulf coast. Area residents needed more advance notice to help with timely evacuations. Hurricane hunters, satellites, and advanced radar have greatly increased the time we have to see and try to predict hurricane paths.

As mentioned previously, the FDA came about because of the pain and suffering of many people. Their system has evolved into one of the toughest to pass among any of our auditing agencies, and for good reason. The FDA makes decisions that affect public safety.

Generally speaking, one should know where they are going before they start any journey. One should know what a building is going to be used for before actually constructing it. For example, a high-tech warehouse would not do a dairy farmer much good, and likewise, a state-of-the-art dairy barn would not do a distribution facility any good. The same can be said about a quality calibration system. Is there a need? What does a company gain by implementing a quality calibration system? Is it worth the time, effort, and expense? Yes, it is! Most governments, industries, and private companies worldwide would agree. The U.S. government regulates to the extent that about 22 percent of all industry in the United States falls under FDA guidelines, which are very specific about the requirements for documented, traceable calibrations (see Chapter 2 for details). All ISO standards have a requirement for calibration when test equipment is involved. The requirements for calibration in other industries such as automotive, airline, nuclear, chemical, and manufacturing are well known throughout their industries. In other words, humankind's need to measure has been around for generations; that need has been and continues to be addressed in a variety of ways.

Humankind understandably turned first to parts of its body and natural surroundings for measuring instruments. Early Babylonian and Egyptian records and the Bible indicate that length was first measured with the forearm, hand, or finger and that time was measured by the periods of the sun, moon, and other heavenly bodies. When it was necessary to compare the capacities of containers such as gourds or clay or metal vessels, they were filled with plant seeds, which were then counted to measure the volume. When means for weighing were invented, seeds and stones served as standards. For instance, the *carat,* still used as a mass unit for gems, was derived from the carob seed.

As societies evolved, measurement units became more complex. The invention of numbering systems and the science of mathematics made it possible to create whole systems of measurement units suited to trade and commerce, land division, taxation, or scientific research. For these more sophisticated uses it was necessary not only to weigh and measure more complex things, it was also necessary to do it accurately time after time and in different places. However, with limited international exchange of goods and communication of ideas, it is not surprising that different systems for the same purpose developed and became established in different parts of the world—even in different parts of a single continent.

The measurement system commonly used in the United States today is nearly the same as that brought by the colonists from England. These measures had their origins in a variety of cultures—Babylonian, Egyptian, Roman, Anglo-Saxon, and Norman-French. The ancient *digit, palm, span,* and *cubit* units evolved into the *inch, foot,* and *yard* through a complicated transformation not yet fully understood.

Roman contributions include the use of the number 12 as a base (our foot is divided into 12 inches) and words from which we derive many of our present measurement unit names. For example, the 12 divisions of the Roman *pes,* or foot, were called *unciae.* Our words *inch* and *ounce* are both derived from that Latin word.

The yard as a measure of length can be traced back to the early Saxon kings. They wore a sash or girdle around the waist that could be removed and used as a convenient measuring device. Thus the word *yard* comes from the Saxon word *gird,* meaning the circumference of a person's waist.

Standardization of the various units and their combinations into a loosely related system of measurement units sometimes occurred in fascinating ways. Tradition holds that King Henry I decreed that the yard should be the distance from the tip of his nose to the end of his thumb. The length of a furlong (or furrow-long) was established by early Tudor rulers as 220 yards. This led Queen Elizabeth I to declare, in the 16th century, that henceforth the traditional Roman mile of 5,000 feet would be replaced by one of 5,280 feet, making the mile exactly eight furlongs and providing a convenient relationship between two previously ill-related measures.

By the 18th century, England had achieved a greater degree of standardization than the continental countries. English units were well-suited to commerce and trade because they had been developed and refined to meet commercial needs. Through colonization and dominance of world commerce during the 17th, 18th, and 19th centuries, the English system of measurement units was spread to and established in many parts of the world, including the American colonies.

However, standards still differed to the extent that they were undesirable for commerce among the 13 colonies. The need for greater uniformity led to clauses in the Articles of Confederation (ratified by the original colonies in 1781) and the Constitution of the United States (ratified in 1790) giving power to the Congress to fix uniform standards for weights and measures. Today, standards supplied to all the states by the National Institute of Standards and Technology (NIST) ensure uniformity throughout the country.[2]

This book is about how to design, implement, maintain, and/or continuously improve a quality calibration system, with all the required documentation, traceability, and known uncertainty for each and every item of test equipment owned and used by any company, large or small. If a business expects to be a player in their market segment, their product must have the quality expected by their customers. This can be accomplished only with test equipment that produces repeatable, accurate, and traceable measurements and/or outputs. Without a quality calibration system in place, this cannot and will not happen. This book will benefit companies that want to implement a program and companies that already have an established program in place.

Some industries have tighter requirements than others on how they treat calibration. Some are more specific about how their standards are read, while being vague about what is needed to meet calibration. Keeping this in mind, this book has been written to meet or exceed these requirements. It does not cost any more to put together a first-class program than it does to put together a ragged-edge one. Both need documentation, records, uncertainty budgets, and written procedures, to name a just a few requirements. And every program needs to continually improve whatever process they have in place on a regular basis.

How can the average calibration technician make an impact within an organization? Are they just another cog in the wheel of industry? No, I believe calibration technicians can have a significant impact on the bottom line of any company and be instrumental in product development, production processes, and customer satisfaction, all through their knowledge of the test equipment that is used in their company. They have a perspective of what works and what doesn't at the ground floor where decisions are made. Their inputs and analysis can make the difference between success and failure of product, process, or manufacturing functions.

NOTES

1. "FDA Protects the Public Health; Ranks High in Public Trust," February 2002. http://www.fda.gov/opacom/factsheets/justthefacts/1fda.html (18 August 2006).

2. "A Brief History of Measurement Systems," 15 September 1999. http://www.slcc.edu/schools/hum_sci/physics/tutor/2210/measurements/history.html (18 August 2006).

2
Requirements and Standards— What They Really Say

Before going into what the current standards and regulations actually state, here is a reminder from times past about measurement practices and how important they really are.

Immersion in water makes the straight seem bent; but reason, thus confused by false appearance, is beautifully restored by measuring, numbering and weighing; these drive vague notions of greater or less or more or heavier right out of the minds of the surveyor, the computer, and the clerk of the scales. Surely it is the better part of thought that relies on measurement and calculation. (Plato, *The Republic*, 360 B.C.E.)[1]

There shall be standard measures of wine, beer, and corn . . . throughout the whole of our kingdom, and a standard width of dyed russet and cloth; and there shall be standard weights also. (Clause 35, *Magna Carta*, 1215)[1]

When you can measure what you are speaking about and express it in numbers, you know something about it; but when you cannot express it in numbers, your knowledge is of a meager and unsatisfactory kind. It may be the beginning of knowledge, but you have scarcely, in your thoughts, advanced to the stage of science. (William Thomson, First Baron Kelvin, GCVO, OM, PC, PRS 26 June 1824–17 December 1907; also known as Lord Kelvin)[1]

All of the following Code of Federal Regulations (CFR) can be located at www.access .gpo.gov/nara/cfr/cfr-table-search.html#page1.[2] Underscore emphasis has been added.

FDA Regulations

The FDA is the federal agency responsible for ensuring that foods are safe, wholesome and sanitary; human and veterinary drugs, biological products, and medical devices are safe and effective; cosmetics are safe; and electronic products that emit radiation are safe. FDA also ensures that these products are honestly, accurately and informatively represented to the public.

21 CFR Part 211

Current Good Manufacturing Practice for Finished Pharmaceuticals—specifies FDA regulations concerning the minimum current good manufacturing practice (cGMP) for preparation of drug products for administration to humans or animals.

TITLE 21—FOOD AND DRUGS

CHAPTER I—FOOD AND DRUG ADMINISTRATION, DEPARTMENT OF HEALTH AND HUMAN SERVICES

Calibration equipment requirements are specified in Section 211.68.

PART 211—CURRENT GOOD MANUFACTURING PRACTICE FOR FINISHED PHARMACEUTICALS

   Subpart D—Equipment

Sec. 211.**68** Automatic, mechanical, and electronic equipment.

   (a) Automatic, mechanical, or electronic equipment or other types of equipment, including computers, or related systems that will perform a function satisfactorily, may be used in the manufacture, processing, packing, and holding of a drug product. If such equipment is so used, it shall be routinely calibrated, inspected, or checked according to a written program designed to assure proper performance. Written records of those calibration checks and inspections shall be maintained.

Laboratory controls procedures are specified in Section 211.160.

PART 211—CURRENT GOOD MANUFACTURING PRACTICE FOR FINISHED PHARMACEUTICALS

   Subpart I—Laboratory Controls

Sec. 211.**160** General requirements.

   (a) The establishment of any specifications, standards, sampling plans, test procedures, or other laboratory control mechanisms required by this subpart, including any change in such specifications, standards, sampling plans, test procedures, or other laboratory control mechanisms, shall be drafted by the appropriate organizational unit and reviewed and approved by the quality control unit. The requirements in this subpart shall be followed and shall be documented at the time of performance. Any deviation from the written specifications, standards, sampling plans, test procedures, or other laboratory control mechanisms shall be recorded and justified.

   (b) Laboratory controls shall include the establishment of scientifically sound and appropriate specifications, standards, sampling plans, and test procedures designed to assure that components, drug product containers, closures, in-process materials, labeling, and drug products conform to appropriate standards of identity, strength, quality, and purity. Laboratory controls shall include:

   (1) Determination of conformance to appropriate written specifications for the acceptance of each lot within each shipment of components, drug product containers, closures, and labeling used in the manufacture, processing, packing, or holding of drug products. The specifications shall include a description of the sampling and testing procedures used. Samples shall be representative and adequately identified. Such procedures shall also require appropriate retesting of any component, drug product container, or closure that is subject to deterioration.

   (2) Determination of conformance to written specifications and a description of sampling and testing procedures for in-process materials. Such samples shall be representative and properly identified.

   (3) Determination of conformance to written descriptions of sampling procedures and appropriate specifications for drug products. Such samples shall be representative and properly identified.

   (4) The calibration of instruments, apparatus, gauges, and recording devices at suitable intervals in accordance with an established written program containing specific directions, schedules, limits for accuracy and precision, and provisions for remedial action in the event accuracy and/or precision limits are not met. Instruments, apparatus, gauges, and recording devices not meeting established specifications shall not be used.

PART 211—CURRENT GOOD MANUFACTURING PRACTICE FOR FINISHED PHARMACEUTICALS

Subpart J—Records and Reports

Sec. 211.**194** Laboratory records.

(a) Laboratory records shall include complete data derived from all tests necessary to assure compliance with established specifications and standards, including examinations and assays, as follows:

(1) A description of the sample received for testing with identification of source (that is, location from where sample was obtained), quantity, lot number or other distinctive code, date sample was taken, and date sample was received for testing.

(2) A statement of each method used in the testing of the sample. The statement shall indicate the location of data that establish that the methods used in the testing of the sample meet proper standards of accuracy and reliability as applied to the product tested. (If the method employed is in the current revision of the United States Pharmacopeia, National Formulary, Association of Official Analytical Chemists, Book of Methods, \1\ or in other recognized standard references, or is detailed in an approved new drug application and the referenced method is not modified, a statement indicating the method and reference will suffice). The suitability of all testing methods used shall be verified under actual conditions of use.

(3) A statement of the weight or measure of sample used for each test, where appropriate.

(4) A complete record of all data secured in the course of each test, including all graphs, charts, and spectra from laboratory instrumentation, properly identified to show the specific component, drug product container, closure, in-process material, or drug product, and lot tested.

(5) A record of all calculations performed in connection with the test, including units of measure, conversion factors, and equivalency factors.

(6) A statement of the results of tests and how the results compare with established standards of identity, strength, quality, and purity for the component, drug product container, closure, in-process material, or drug product tested.

(7) The initials or signature of the person who performs each test and the date(s) the tests were performed.

(8) The initials or signature of a second person showing that the original records have been reviewed for accuracy, completeness, and compliance with established standards.

(b) Complete records shall be maintained of any modification of an established method employed in testing. Such records shall include the reason for the modification and data to verify that the modification produced results that are at least as accurate and reliable for the material being tested as the established method.

(c) Complete records shall be maintained of any testing and standardization of laboratory reference standards, reagents, and standard solutions.

(d) Complete records shall be maintained of the periodic calibration of laboratory instruments, apparatus, gauges, and recording devices required by Sec. 211.160(b)(4).

(e) Complete records shall be maintained of all stability testing performed in accordance with Sec. 211.166.

21 CFR Part 820

Quality System Regulation—specifies FDA requirements that govern the methods used in, and the facilities and controls used for, the design, manufacture, packaging, labeling, storage, installation, and servicing of all finished devices intended for human use. Current good manufacturing practice (cGMP) requirements are set forth in this regulation.

Calibration requirements are specified in Sec. 820.72.

PART 820—QUALITY SYSTEM REGULATION

Subpart G—Production and Process Controls

Sec. 820.**72** Inspection, measuring, and test equipment.

(a) Control of inspection, measuring, and test equipment. Each manufacturer shall ensure that all inspection, measuring, and test equipment, including mechanical, automated, or electronic inspection and test equipment, is suitable for its intended purposes and is capable of producing valid results. Each manufacturer shall establish and maintain procedures to ensure that equipment is routinely calibrated, inspected, checked, and maintained. The procedures shall include provisions for handling, preservation, and storage of equipment, so that its accuracy and fitness for use are maintained. These activities shall be documented.

(b) Calibration. Calibration procedures shall include specific directions and limits for accuracy and precision. When accuracy and precision limits are not met, there shall be provisions for remedial action to reestablish the limits and to evaluate whether there was any adverse effect on the device's quality. These activities shall be documented.

(1) Calibration standards. Calibration standards used for inspection, measuring, and test equipment shall be traceable to national or international standards. If national or international standards are not practical or available, the manufacturer shall use an independent reproducible standard. If no applicable standard exists, the manufacturer shall establish and maintain an in-house standard.

(2) Calibration records. The equipment identification, calibration dates, the individual performing each calibration, and the next calibration date shall be documented. These records shall be displayed on or near each piece of equipment or shall be readily available to the personnel using such equipment and to the individuals responsible for calibrating the equipment.

21 CFR Part 58

Good Laboratory Practices For Nonclinical Laboratory Studies—describes good laboratory practices (GLP) for conducting nonclinical laboratory studies that support or are intended to support applications for research or marketing permits for products regulated by the FDA, including food and color additives, animal food additives, human and animal drugs, medical devices for human use, biological products, and electronic products.

Maintenance and calibration of equipment are specified in Section 58.63.

PART 58—GOOD LABORATORY PRACTICE FOR NONCLINICAL LABORATORY STUDIES

Subpart D—Equipment

Sec. 58.**63** Maintenance and calibration of equipment.

(a) Equipment shall be adequately inspected, cleaned, and maintained. Equipment used for the generation, measurement, or assessment of data shall be adequately tested, calibrated and/or standardized.

(b) The written standard operating procedures required under Sec. 58.81(b)(11) shall set forth in sufficient detail the methods, materials, and schedules to be used in the routine inspection, cleaning, maintenance, testing, calibration, and/or standardization of equipment, and shall specify, when appropriate, remedial action to be taken in the event of failure or malfunction of equipment. The written standard operating procedures shall designate the person responsible for the performance of each operation.

(c) Written records shall be maintained of all inspection, maintenance, testing, calibrating, and/or standardizing operations. These records, containing the date of the operation, shall describe whether the maintenance operations were routine and followed the written standard operating procedures. Written records shall be kept of nonroutine repairs performed on equipment as a result of failure and malfunction. Such records shall document the nature of the defect, how and when the defect was discovered, and any remedial action taken in response to the defect.

Standard operating procedures are specified in Section 58.81.

PART 58—GOOD LABORATORY PRACTICE FOR NONCLINICAL LABORATORY STUDIES

Subpart E—Testing Facilities Operation

Sec. 58.**81** Standard operating procedures.

(a) A testing facility shall have standard operating procedures in writing setting forth nonclinical laboratory study methods that management is satisfied are adequate to insure the quality and integrity of the data generated in the course of a study. All deviations in a study from standard operating procedures shall be authorized by the study director and shall be documented in the raw data. Significant changes in established standard operating procedures shall be properly authorized in writing by management.

(b) Standard operating procedures shall be established for, but not limited to, the following:

(1) Animal room preparation.

(2) Animal care.

(3) Receipt, identification, storage, handling, mixing, and method of sampling of the test and control articles.

(4) Test system observations.

(5) Laboratory tests.

(6) Handling of animals found moribund or dead during study.

(7) Necropsy of animals or postmortem examination of animals.

(8) Collection and identification of specimens.

(9) Histopathology.

(10) Data handling, storage, and retrieval.

(11) Maintenance and calibration of equipment.

(12) Transfer, proper placement, and identification of animals.

(c) Each laboratory area shall have immediately available laboratory manuals and standard operating procedures relative to the laboratory procedures being performed. Published literature may be used as a supplement to standard operating procedures.

(d) A historical file of standard operating procedures, and all revisions thereof, including the dates of such revisions, shall be maintained.

21 CFR Part 110

Current Good Manufacturing Practice in Manufacturing, Packing, or Holding Human Food Equipment and utensil maintenance is specified in Section 110.40.

PART 110—CURRENT GOOD MANUFACTURING PRACTICE IN MANUFACTURING, PACKING, OR HOLDING HUMAN FOOD

Subpart C—Equipment

Sec. 110.**40** Equipment and utensils.

(a) All plant equipment and utensils shall be so designed and of such material and workmanship as to be adequately cleanable, and shall be properly maintained. The design, construction, and use of equipment and utensils shall preclude the adulteration of food with lubricants, fuel, metal fragments, contaminated water, or any other contaminants. All equipment should be so installed and maintained as to facilitate the cleaning of the equipment and of all adjacent spaces. Food-contact surfaces shall be corrosion-resistant when in contact with food. They shall be made of nontoxic materials and designed to withstand the environment of their intended use and the action of food, and, if applicable, cleaning compounds and sanitizing agents. Food-contact surfaces shall be maintained to protect food from being contaminated by any source, including unlawful indirect food additives.

(b) Seams on food-contact surfaces shall be smoothly bonded or maintained so as to minimize accumulation of food particles, dirt, and organic matter and thus minimize the opportunity for growth of microorganisms.

(c) Equipment that is in the manufacturing or food-handling area and that does not come into contact with food shall be so constructed that it can be kept in a clean condition.

(d) Holding, conveying, and manufacturing systems, including gravimetric, pneumatic, closed, and automated systems, shall be of a design and construction that enables them to be maintained in an appropriate sanitary condition.

(e) Each freezer and cold storage compartment used to store and hold food capable of supporting growth of microorganisms shall be fitted with an indicating thermometer, temperature-measuring device, or temperature-recording device so installed as to show the temperature accurately within the compartment, and should be fitted with an automatic control for regulating temperature or with an automatic alarm system to indicate a significant temperature change in a manual operation.

(f) Instruments and controls used for measuring, regulating, or recording temperatures, pH, acidity, water activity, or other conditions that control or prevent the growth of undesirable microorganisms in food shall be accurate and adequately maintained, and adequate in number for their designated uses.

(g) Compressed air or other gases mechanically introduced into food or used to clean food-contact surfaces or equipment shall be treated in such a way that food is not contaminated with unlawful indirect food additives.

21 CFR Part 606

Current Good Manufacturing Practice for Blood and Blood Components—specifies FDA regulations concerning the current good manufacturing practice (cGMP) for the collection, processing, and storage of blood and blood components.

Calibration of equipment is specified in Section 606.60.

PART 606—CURRENT GOOD MANUFACTURING PRACTICE FOR BLOOD AND BLOOD COMPONENTS

Subpart D—Equipment

Sec. 606.60 Equipment.

(a) Equipment used in the collection, processing, compatibility testing, storage, and distribution of blood and blood components shall be maintained in a clean and orderly manner and located so as to facilitate cleaning and maintenance. The equipment shall be observed, standardized, and calibrated on a regularly scheduled basis as prescribed in the Standard Operating Procedures Manual and shall perform in the manner for which it was designed so as to assure compliance with the official requirements prescribed in this chapter for blood and blood products.

(b) Equipment shall be observed, standardized, and calibrated with at least the following frequency, include but are not limited to: (refer to the CFR for this table)

(c) Equipment employed in the sterilization of materials used in blood collection or for disposition of contaminated products shall be designed, maintained, and utilized to ensure the destruction of contaminating microorganisms. The effectiveness of the sterilization procedure shall be no less than that achieved by an attained temperature of 121.5 °C (251 °F) maintained for 20 minutes by saturated steam or by an attained temperature of 170 °C (338 °F) maintained for 2 hours with dry heat.

Standard operating procedures are specified in Section 606.100.

PART 606—CURRENT GOOD MANUFACTURING PRACTICE FOR BLOOD AND BLOOD COMPONENTS

Subpart F—Production and Process Controls

Sec. 606.**100** Standard operating procedures.

(a) In all instances, except clinical investigations, standard operating procedures shall comply with published additional standards in part 640 of this chapter for the products being processed; except that, references in part 640 relating to licenses, licensed establishments, and submission of material or data to or approval by the Director, Center for Biologics Evaluation and Research, are not applicable to establishments not subject to licensure under section 351 of the Public Health Service Act.

(b) Written standard operating procedures shall be maintained and shall include all steps to be followed in the collection, processing, compatibility testing, storage, and distribution of blood and blood components for transfusion and further manufacturing purposes. Such procedures shall be available to the personnel for use in the areas where the procedures are performed. The written standard operating procedures shall include, but are not limited to, descriptions of the following, when applicable:

(1) Criteria used to determine donor suitability, including acceptable medical history criteria.

(2) Methods of performing donor qualifying tests and measurements, including minimum and maximum values for a test or procedure when a factor in determining acceptability.

(3) Solutions and methods used to prepare the site of phlebotomy to give maximum assurance of a sterile container of blood.

(4) Method of accurately relating the product(s) to the donor.

(5) Blood collection procedure, including in-process precautions taken to measure accurately the quantity of blood removed from the donor.

(6) Methods of component preparation, including any time restrictions for specific steps in processing.

(7) All tests and repeat tests performed on blood and blood components during manufacturing.

(8) Pretransfusion testing, where applicable, including precautions to be taken to identify accurately the recipient blood samples and cross-matched donor units.

(9) Procedures for investigating adverse donor and recipient reactions.

(10) Storage temperatures and methods of controlling storage temperatures for all blood products and reagents as prescribed in Secs. 600.15 and 610.53 of this chapter.

(11) Length of expiration dates, if any, assigned for all final products as prescribed in Sec. 610.53 of this chapter.

(12) Criteria for determining whether returned blood is suitable for reissue.

(13) Procedures used for relating a unit of blood or blood component from the donor to its final disposition.

(14) Quality control procedures for supplies and reagents employed in blood collection, processing and pretransfusion testing.

(15) Schedules and procedures for equipment maintenance and calibration.

(16) Labeling procedures, including safeguards to avoid labeling mixups.

(17) Procedures of plasmapheresis, plateletpheresis, and leukapheresis, if performed, including precautions to be taken to ensure reinfusion of a donor's own cells.

(18) Procedures for preparing recovered plasma, if performed, including details of separation, pooling, labeling, storage, and distribution.

(19) Procedures in accordance with Sec. 610.46 of this chapter to look at prior donations of Whole Blood, blood components, Source Plasma, and Source Leukocytes from a donor who has donated blood and subsequently tests repeatedly reactive for antibody to human immunodeficiency

virus (HIV) or otherwise is determined to be unsuitable when tested in accordance with Sec. 610.45 of this chapter. Procedures to quarantine in-house Whole Blood, blood components, Source Plasma, and Source Leukocytes intended for further manufacture into injectable products that were obtained from such donors; procedures to notify consignees regarding the need to quarantine such products; procedures to determine the suitability for release of such products from quarantine; procedures to notify consignees of Whole Blood, blood components, Source Plasma, and Source Leukocytes from such donors of the results of the antibody testing of such donors; and procedures in accordance with Sec. 610.47 of this chapter to notify attending physicians so that transfusion recipients are informed that they may have received Whole Blood and blood components at increased risk for transmitting human immunodeficiency virus.

(c) All records pertinent to the lot or unit maintained pursuant to these regulations shall be reviewed before the release or distribution of a lot or unit of final product. The review or portions of the review may be performed at appropriate periods during or after blood collecting, processing, compatibility testing, and storing. A thorough investigation, including the conclusions and followup, of any unexplained discrepancy or the failure of a lot or unit to meet any of its specifications shall be made and recorded.

(d) In addition to the requirements of this subpart and in conformity with this section, any facility may utilize current standard operating procedures such as the manuals of the organizations, as long as such specific procedures are consistent with, and at least as stringent as, the requirements contained in this part.

(1) American Association of Blood Banks.

(2) American National Red Cross.

(3) Other organizations or individual blood banks, subject to approval by the Director, Center for Biologics Evaluation and Research.

Recordkeeping is specified in Section 606.160.

PART 606—CURRENT GOOD MANUFACTURING PRACTICE FOR BLOOD AND BLOOD COMPONENTS

Subpart I—Records and Reports

Sec. 606.**160** Records.

(a)(1) Records shall be maintained concurrently with the performance of each significant step in the collection, processing, compatibility testing, storage and distribution of each unit of blood and blood components so that all steps can be clearly traced. All records shall be legible and indelible, and shall identify the person performing the work, include dates of the various entries, show test results as well as the interpretation of the results, show the expiration date assigned to specific products, and be as detailed as necessary to provide a complete history of the work performed.

(2) Appropriate records shall be available from which to determine lot numbers of supplies and reagents used for specific lots or units of the final product.

(b) Records shall be maintained that include, but are not limited to, the following when applicable:

(1) Donor records:

(i) Donor selection, including medical interview and examination and where applicable, informed consent.

(ii) Permanent and temporary deferrals for health reasons including reason(s) for deferral.

(iii) Donor adverse reaction complaints and reports, including results of all investigations and followup.

(iv) Therapeutic bleedings, including signed requests from attending physicians, the donor's disease and disposition of units.

(v) Immunization, including informed consent, identification of the antigen, dosage, and route of administration.

(vi) Blood collection, including identification of the phlebotomist.

(vii) Records to relate the donor with the unit number of each previous donation from that donor.

(viii) Records of quarantine, notification, testing, and disposition performed pursuant to Secs. 610.46 and 610.47 of this chapter.

(2) Processing records:

(i) Blood processing, including results and interpretation of all tests and retests.

(ii) Component preparation, including all relevant dates and times.

(iii) Separation and pooling of recovered plasma.

(iv) Centrifugation and pooling of source plasma.

(v) Labeling, including initials of the person(s) performing the procedure.

(3) Storage and distribution records:

(i) Distribution and disposition, as appropriate, of blood and blood products.

(ii) Visual inspection of whole blood and red blood cells during storage and immediately before distribution.

(iii) Storage temperature, including initialed temperature recorder charts.

(iv) Reissue, including records of proper temperature maintenance.

(v) Emergency release of blood, including signature of requesting physician obtained before or after release.

(4) Compatibility test records:

(i) Results of all compatibility tests, including cross-matching, testing of patient samples, antibody screening, and identification.

(ii) Results of confirmatory testing.

(5) Quality control records:

(i) Calibration and standardization of equipment.

(ii) Performance checks of equipment and reagents.

(iii) Periodic check on sterile technique.

(iv) Periodic tests of capacity of shipping containers to maintain proper temperature in transit.

(v) Proficiency test results.

(6) Transfusion reaction reports and complaints, including records of investigations and followup.

(7) General records:

(i) Sterilization of supplies and reagents prepared within the facility, including date, time interval, temperature, and mode.

(ii) Responsible personnel.

(iii) Errors and accidents.

(iv) Maintenance records for equipment and general physical plant.

(v) Supplies and reagents, including name of manufacturer or supplier, lot numbers, expiration date, and date of receipt.

(vi) Disposition of rejected supplies and reagents used in the collection, processing, and compatibility testing of blood and blood components.

(c) A donor number shall be assigned to each accepted donor, which relates the unit of blood collected to that donor, to his medical record, to any component or blood product from that donor's unit of blood, and to all records describing the history and ultimate disposition of these products.

(d) Records shall be retained for such interval beyond the expiration date for the blood or blood component as necessary to facilitate the reporting of any unfavorable clinical reactions. The retention period shall be no less than 5 years after the records of processing have been completed or 6 months after the latest expiration date for the individual product, whichever is a later date. When there is no expiration date, records shall be retained indefinitely.

(e) A record shall be available from which unsuitable donors may be identified so that products from such individuals will not be distributed.

According to the cGMP regulations, equipment must be qualified, calibrated, cleaned, and maintained to prevent contamination and mix-ups (§§ 211.63, 211.67, 211.68). Note that the cGMP regulations require a higher standard for calibration and maintenance than most generic quality system models. The cGMP regulations place as much emphasis on process equipment as on testing equipment (§ 211.42(b)), while most quality systems focus only on testing equipment.[4]

TITLE 14—Aeronautics and Space

CHAPTER I—FEDERAL AVIATION ADMINISTRATION, DEPARTMENT OF TRANSPORTATION

SUBCHAPTER G—AIR CARRIERS AND OPERATORS FOR COMPENSATION OR HIRE: CERTIFICATION AND OPERATIONS

PART 121—OPERATING REQUIREMENTS: DOMESTIC, FLAG, AND SUPPLEMENTAL OPERATIONS

Subpart L—Maintenance, Preventive Maintenance, and Alterations

Sec. 121.369 Manual requirements.

(a) The certificate holder shall put in its manual a chart or description of the certificate holder's organization required by Sec. 121.365 and a list of persons with whom it has arranged for the performance of any of its required inspections, other maintenance, preventive maintenance, or alterations, including a general description of that work.

(b) The certificate holder's manual must contain the programs required by Sec. 121.367 that must be followed in performing maintenance, preventive maintenance, and alterations of that certificate holder's airplanes, including airframes, aircraft engines, propellers, appliances, emergency equipment, and parts thereof, and must include at least the following:

(1) The method of performing routine and non-routine maintenance (other than required inspections), preventive maintenance, and alterations.

(2) A designation of the items of maintenance and alteration that must be inspected (required inspections), including at least those that could result in a failure, malfunction, or defect endangering the safe operation of the aircraft, if not performed properly or if improper parts or materials are used.

(3) The method of performing required inspections and a designation by occupational title of personnel authorized to perform each required inspection.

(4) Procedures for the re-inspection of work performed pursuant to previous required inspection findings (buy-back procedures).

(5) Procedures, standards, and limits necessary for required inspections and acceptance or rejection of the items required to be inspected and for periodic inspection and calibration of precision tools, measuring devices, and test equipment.

(6) Procedures to ensure that all required inspections are performed.

(7) Instructions to prevent any person who performs any item of work from performing any required inspection of that work.

(8) Instructions and procedures to prevent any decision of an inspector, regarding any required inspection from being countermanded by persons other than supervisory personnel of the inspection unit, or a person at that level of administrative control that has overall responsibility for the management of both the required inspection functions and the other maintenance, preventive maintenance, and alterations functions.

(9) Procedures to ensure that required inspections, other maintenance, preventive maintenance, and alterations that are not completed as a result of shift changes or similar work interruptions are properly completed before the aircraft is released to service.

(c) The certificate holder must set forth in its manual a suitable system (which may include a coded system) that provides for preservation and retrieval of information in a manner acceptable to the Administrator and that provides—

(1) A description (or reference to data acceptable to the Administrator) of the work performed;

(2) The name of the person performing the work if the work is performed by a person outside the organization of the certificate holder; and

(3) The name or other positive identification of the individual approving the work.

PART 145—REPAIR STATIONS

Subpart E—Operating Rules

Sec. 145.211 Quality control system.

(a) A certificated repair station must establish and maintain a quality control system acceptable to the FAA that ensures the airworthiness of the articles on which the repair station or any of its contractors performs maintenance, preventive maintenance, or alterations.

(b) Repair station personnel must follow the quality control system when performing maintenance, preventive maintenance, or alterations under the repair station certificate and operations specifications.

(c) A certificated repair station must prepare and keep current a quality control manual in a format acceptable to the FAA that includes the following:

(1) A description of the system and procedures used for—

(i) Inspecting incoming raw materials to ensure acceptable quality;

(ii) Performing preliminary inspection of all articles that are maintained;

(iii) Inspecting all articles that have been involved in an accident for hidden damage before maintenance, preventive maintenance, or alteration is performed;

(iv) Establishing and maintaining proficiency of inspection personnel;

(v) Establishing and maintaining current technical data for maintaining articles;

(vi) Qualifying and surveilling non-certificated persons who perform maintenance, prevention maintenance, or alterations for the repair station;

(vii) Performing final inspection and return to service of maintained articles;

(viii) Calibrating measuring and test equipment used in maintaining articles, including the intervals at which the equipment will be calibrated; and

(ix) Taking corrective action on deficiencies;

(2) References, where applicable, to the manufacturer's inspection standards for a particular article, including reference to any data specified by that manufacturer;

(3) A sample of the inspection and maintenance forms and instructions for completing such forms or a reference to a separate forms manual; and

(4) Procedures for revising the quality control manual required under this section and notifying the certificate holding district office of the revisions, including how often the certificate holding district office will be notified of revisions.

(d) A certificated repair station must notify its certificate holding district office of revisions to its quality control manual.

ISO/TS 16949: 2002, Quality Management Systems–Particular Requirements for the Application of ISO 9001: 2000 for Automotive Production and Relevant Service Part Organizations.

For the benefit of readers who are not familiar with what is meant by *the "automotive industry"*, it can be defined as companies eligible to adopt ISO/TS 16949:2002. This is explained by Graham Hills as the manufacturers of finished automotive products and:

". . . suppliers that make or fabricate production materials, production or service parts or production part assemblies. It also applies to specific service-oriented suppliers—heat treating, welding, painting, plating or other finishing services. The customers for these types of suppliers must be manufacturers of automobiles, trucks (heavy, medium and light duty), buses, or motorcycles. [It] does not apply to manufacturing suppliers for off-highway, agricultural, or mining OEMs. It also doesn't apply to service-oriented suppliers offering distribution, warehousing, sorting or non-value-added services. Nor does it apply to aftermarket parts manufacturers."[3]

ANSI/ISO/ASQ Q9001-2000 discusses calibration in Section 7.6 *Control of measuring and monitoring devices*. If a device makes a measurement that provides evidence of conformity to requirements, or if it is necessary for the process and if it must be ensured that the measurement results are valid, then certain things must be done.

- The test equipment must be regularly calibrated against measurement standards traceable to (the SI through) international or national measurement standards, or other accepted standards if there is no such traceability.
- It must be adjusted, if indicated by calibration results.
- It must be identified in a manner that allows the user to determine the calibration status.
- If the item has any controls or adjustments that would invalidate the calibration, then it must be protected from access by the user.
- It must be protected from damage, and from environmental conditions that could damage or degrade it, whenever it is not in use.
- If a calibration shows that the *as received* condition was out of tolerance, the organization must assess the condition and its impact on any product, take appropriate action, and keep appropriate records including results of the calibration.
- If computer software is used to monitor and measure requirements, then the organization must prove that the software operates as intended and produces valid results before placing it in regular use.[4]

NOTES

1. "BIPM—Bureau International des Poids et Measures. Metrology and World Metrology Day," June 2006. http://www.bipm.org/en/practical_info/faq/welcome.html# (18 August 2006). "Magna Carta," 21 August 2006. http://en.wikipedia.org/wiki/Magna_Carta (21 August 2006). Magna Carta (Latin for "Great Charter," literally "Great Paper"), also called Magna Carta Libertatum ("Great Charter of Freedoms"), was an English charter originally issued in 1215. Magna Carta is the most significant early influence on the long historical process that led to the rule of constitutional law today. Magna Carta was originally created because of disagreements between the pope, King John and his English barons about the rights of the king. Magna Carta required the king to renounce certain rights, respect certain legal procedures, and accept that *the will of the king could be bound by law*. There are a number of popular misconceptions about Magna Carta, such as that it was the first document to limit the power of an English king by law (it was not the first, and was partly based on the Charter of Liberties); that it in practice limited the power of the king (it mostly did not in the Middle Ages); and that it is a single static document (it is a variety of documents referred to under a common name). Magna Carta was renewed throughout the Middle Ages, and further during the Tudor and Stuart periods and the 17th and 18th centuries. By the early 19th century most clauses had been repealed from English law. The influence of Magna Carta outside England can be seen in the United States Constitution and Bill of Rights. Indeed, just about every common law country with a constitution has been influenced by Magna Carta, making it one of the most important legal documents in the history of democracy.

2. "Guidance for Industry, Quality Systems Approach to Pharmaceutical Current Good Manufacturing Practice Regulations, Draft Guidance," September 2004. http://www.fda.gov/cder/guidance/6452dft.pdf (18 August 2006).

3. Jay L. Bucher, *The Metrology Handbook* (Milwaukee: ASQ Quality Press, 2004), 124.

4. Bucher, *The Metrology Handbook,* 114.

Part II

The Basics of a Quality Calibration System

3

The Basics

One should keep in mind that calibration is a process (procedures, records, analysis, and communication when test equipment is out of tolerance, and so on), not an event. Each calibration is an action that takes place once for a particular item, but the action continues to be repeated. The data can be used for many functions, including: calibration interval analysis, alert/action procedures, statistical analysis, to see if processes are or are not functioning properly, and limited calibrations, just to name a few. This is all part of a mind-set of calibration practitioners from technicians to supervisors to management. Calibration is a critical and necessary process that could make the difference between life and death, profit and bankruptcy, and good or bad products.

The basic premise and foundation of a quality calibration system is to "*say what you do, do what you say, record what you did, check the results, and act on the difference.*" Let's break these down into simple terms that will be discussed in great detail in the following chapters.[1]

Say what you do means write in detail how to do your job. This includes calibration procedures, standard operating procedures (SOPs), protocols, work instruction, and/or work cards. Companies use different names for their procedures, but they must have procedures in place and available to use. What those procedures are is specifically described in various regulations, recommended practices, or procedures. Chapter 4 goes into detail on calibration procedures.

Do what you say means follow the documented procedures or instructions every time you calibrate or perform a function that follows specific written instructions. The requirement for following written procedures is spelled out everywhere you look. It only makes sense that if you are required to have written procedures, then you must follow them. Why is this so important? Simply put, if repeatable, reliable, and traceable calibrations are to be performed, the calibration must be performed the same way each and every time. This can only happen when a specific set of instructions are followed. Deviation from those instructions does not allow for repeatable calibrations.

In order to compare historical data on a piece of test equipment, one must be able to compare apples to apples and oranges to oranges, not apples to oranges. If different standards having different tolerances—different standards may be used for identical calibrations, but during substitution of standards, their specifications must be equal to or better than those of the substituted standard(s) that had lesser tolerances—the resulting readings could not be compared. (See Chapter 7 for more details.)

Record what you did means that you must record the results of your measurements and adjustments, including what your standard(s) read or indicated, as well as the test instrument being calibrated, both before and after any adjustments. This is also spelled out in different regulations. The specifics and how to conform to those requirements can be found in Chapter 5.

Check the results means that you must make certain the test equipment meets the tolerances, accuracies, or upper/lower limits specified in your procedures or instructions. Depending on your industry, you may only have to ensure that test equipment meets a set of tolerances or specifications. However, most requirements in the calibration community specify that the data be collected and stored. This requirement has many advantages besides just ensuring that a tolerance is met. Data cannot be retrieved for comparison, statistical analysis, or used for calibration interval analysis if the data is not available. Just saying that a tolerance was met is no longer the preferred way of performing calibrations. Chapter 6 goes into detail about what should happen when a piece of test equipment does not meet its stated tolerances.

Act on the difference means that if the test equipment is out of tolerance, does not meet the specified accuracies, or exceeds the upper/lower test limits written in your procedures, you are required to inform the user/owner of the equipment because they may have to re-evaluate manufactured goods, change a process, or recall a product. The worst-case scenario is that previously calibrated equipment that used that particular standard must be recalled and recalibrated, along with any products that used any of the suspect test equipment. This could become costly and very time consuming if procedures are not in place for reverse traceability (see Chapter 8).

As stated in the first sentence of this chapter, the previous five steps form the foundation for a quality calibration system. However, one must build on that foundation to ensure that it works properly. Part of any system is the function used to update procedures, records, and quality system. Documentation control is critical to having the correct, up-to-date procedure in place when it is needed, a calibration technician who is trained on that procedure and any changes or modifications to it. All of this requires controls, documentation, and training.

According to ANSI/ISO/IEC 17025-2005, Chapter 4.3, *Document control:* "The laboratory shall establish and maintain procedures to control all documents . . . all documents issued . . . shall be reviewed and approved. A master list . . . identifying the current revision status and distribution of documents in the quality system shall be established. Changes to documents shall be reviewed and approved."[2] Q9001-2000, Chapter 4.2.3, *Control of documents* states: "Documents required by the quality management system shall be controlled. A documented procedure shall be established to define the controls needed to approve documents . . . review and update . . . ensure that changes are identified . . . that relevant versions are available . . . . to prevent the unintended use of obsolete documents." Z540 states in Chapter 5.2: "The quality manual and related documentation shall also contain . . . procedures for control and maintenance of documentation."[3]

One of the most important parts of a quality system should be the document control procedures. How do you control who makes changes, how new documents and/or changes to old documents get posted, and when the users are notified that changes have been made? There are software packages available that can assist in controlling an organization's document system, but a small business may not be able to afford those packages. Following is a brief overview of what can be done with the resources an organization may already be using.

Each of your controlled documents (procedures, records, certificates, and so on) should have a unique identification (many systems use a number, often known as a control number) as well as a revision number. A master list with all of this information should be available for anyone to see what the current documents are within their quality system. This list should also include the revision date of the document. It needs to be updated every time changes are made and approved for your documents. A simple spreadsheet or word processor document can fulfill this requirement, as long as it is updated and maintained. And it has to be available to the technicians who use the various procedures, records, and certificates. This isn't a difficult problem in a small group where only a few people work with the various documents. But if an organization has two or three shifts, different locations, or off-site calibration responsibilities, the opportunity for using out-of-date or incorrect documents could easily become a problem. Notification of changes and training, when applicable, should be documented as a form of keeping everyone informed and up to date. Part of an organization's training program should include when and how to inform and train its staff when there are changes to their documents. Some organizations ensure training and/or notification of changes has occurred before they allow the latest revisions to be posted. In some systems, the new revision must be posted in order for the user to have access to the documents. Another approach would be to maintain all of your quality documentation electronically via an intranet. In this system, any printed documents would be invalid. This process ensures that only the latest and greatest procedures are available to all who need them. No matter which way an organization's system works, it is vital that everyone involved be informed and trained when changes are made and that only the latest revisions are available for their use.

As a minimum, for each of its controlled documents an organization's master list should have a unique identification, document name, revision or edition number, and revision date. In addition, it is helpful to include the name of the approver or approval authority and something stating that only the revisions listed should be used. Archiving a copy of previous revisions can have benefits, but those revisions must be kept in a location where they cannot be readily accessed for use by staff. Some systems use black lines in their borders to indicate where changes have been made; others annotate the changes in a reference section at the end of the document; and still others refer to comparisons of archived documents as the only reference to changes. No matter which system or combination of systems is used, they only need to meet the quality system requirements that have been set for the organization.

NOTES

1. Jay L. Bucher, *The Metrology Handbook* (Milwaukee: ASQ Quality Press, 2004), 19.

2. ANSI/ISO/IEC, *ANSI/ISO/IEC 17025-2005: General requirements for the competence of testing and calibration laboratories* (Milwaukee: ASQ Quality Press, 2005).

3. Bucher, *The Metrology Handbook,* 37.

4

Calibration Procedures

By definition, a *calibration procedure* is a controlled document that provides a validated method for evaluating and verifying the essential performance characteristics, specifications, or tolerances for test equipment. A calibration procedure documents one method of verifying the actual performance of the item being calibrated against its performance specifications. It provides a list of recommended calibration standards to use for the calibration, a means to record quantitative performance data both before (called *As Found* readings) and after (called *As Left* readings) adjustments, and information sufficient to determine if the unit under calibration is operating within the necessary performance specifications.

It matters little what name is given to this part of your quality calibration system. Calibration procedure, standard operating procedure (SOP), protocol, work instructions, work cards, or any number of other monikers. All of these terms are used to describe how to calibrate a particular piece of test equipment.

There should be three parts to this system. The first part is the procedure itself—how it is written and what it contains. The second part is the control system used to ensure that changes, updates, and modifications to the calibration procedure are implemented. It is not enough to have the procedures if there is no way to continuously improve them. Nothing is written in stone, most certainly not calibration procedures. Changes are made every day to the standards used because of updates from the manufacturer, new modifications to old items, and many other reasons. There must be a process in place to make the changes, update the written procedures that the calibration technicians use, and inform and train technicians on the improvements and/or changes. Also, when errors are found, the procedure needs to be changed, as well.

The third part is the most critical. The calibration procedure must be used each and every time a calibration is performed. This is important enough to repeat: *The calibration procedure must be used each and every time a calibration is performed.* Every standard and regulation calls for the use of calibration procedures. The standards and regulations do not say that calibration procedures have to be made available, but that they will be used.

This is where the real world collides with standards and regulations. Everyone who has performed numerous calibrations on a particular piece or type of test equipment over many years will be thinking to themselves: "I've done this a million times—why do I need to look at the same old procedure?" Is it possible the procedure has changed? Is it possible the standards used have changed? Is it possible the item being calibrated has changed? These are all possibilities. But the question in most calibration technician's minds is, "If I know what I am doing, why do I have to reread a procedure that I know by heart?"

The nice part of ISO and other standards is that nobody has to memorize or know many things. The one thing that users must know is where to find the information when needed. The information on performing any particular calibration is written in a calibration procedure. That calibration procedure must be with the calibration technicians when they are performing the calibration. In order to perform the calibration unsupervised, there must be documentation stating that the calibration technician is proficient in doing that task. If the calibration technician is signed off as being able to perform unsupervised calibrations, that is step one. Step two is to have all required tools and documents available to perform the task. One of those required documents is the calibration procedure. It must be in the general area with the technicians when they are performing the calibration. An auditor does not expect the experienced calibration technician to constantly refer to the procedure each and every time a calibration is performed. However, they are expected to know where to find something within the calibration procedure if there are any questions about what is being accomplished. This is where experienced calibration technicians know what is in the procedure, can easily turn to the correct page, and can show they are performing the step as written in the procedure. If they cannot do this, they should be following the procedure step by step on every calibration.

Humans are creatures of habit. They get into a groove and don't want to make changes. This is part of why it is critical to ensure that every time a change is made to a calibration procedure, it is documented and that every calibration technician is trained on what the change is and how to correctly perform the new procedure.

A calibration procedure is only as good as what is written on its pages and the calibration technician who follows it. In order to have repeatable, reliable, accurate measurements with test equipment, each item must be calibrated the same way. This can be accomplished only by using and following validated calibration procedures each and every time a calibration is performed.

All quality systems that address calibration require written instructions for the calibration of test equipment. Under the quality system, this is the *say what you do* portion, which means how to do the job must be written down in detail (this includes calibration procedures, SOPs, protocols, work instruction, work cards, and so on). Why do formal instructions or procedures need to be followed? In order to get consistent results, step-by-step instructions must be followed each and every time calibrations are performed.

The following is paraphrased from different sources focusing on written calibration procedures: ANSI/ISO/ASQ Q10012-2003, *Measurement management systems—Requirements for measurement processes and measuring equipment;* NCSL International's RP-6, *Calibration Control Systems for the Biomedical and Pharmaceutical Industry;* and ANSI/ASQ M1-1996, *American National Standard for Calibration Systems.* Another source is NCSL International's RP-3, *Calibration Procedures.*

ANSI/ISO/ASQ Q10012-2003, Chapter 6.2.1, states, "Measurement management system procedures shall be documented to the extent necessary and validated to ensure the proper implementation, their consistency of application, and the validity of measurement results. New procedures or changes to documented procedures shall be authorized and controlled. Procedures shall be current, available and provided when required."

NCSLI RP-6 states in Chapter 5.9, *Calibration Procedures,* "Documentation should be provided containing sufficient information for the calibration of measurement equipment." The requirements in 5.9.1 are:

- Source (The calibration procedure may be . . . prepared internally, by another agency, by the manufacturer, or by a composite of the three.)
- Completeness (The procedure should contain sufficient instruction and information to enable qualified personnel to perform the calibration.)
- Approval (All procedures should be approved and controlled. . . . Evidence should be displayed on the document.)
- Software (Any used instead of an actual procedure should follow the computer software recommendation for control.)

Under *Format* in Chapter 5.9.2, internal procedures should include:

- Performance requirements (device description, manufacture, type or model number, environmental conditions, specifications . . .)
- Measurement standards (generic description of measurement standards and performance requirements, accuracy ratio and/or uncertainty, and any auxiliary tools)
- Preliminary operations (any safety or handling requirements, cleaning prerequisites, reminders, or operational checks)
- Calibration process (the detailed set of instructions for process verification in well-defined segments . . . upper and lower tolerance limits, and required further instructions)
- Calibration results (performance results data sheet or form to record the calibration data when required)
- Closing operations (any labeling, calibration safeguards, and material-removal requirements to prevent contamination of product)
- Storage and handling (. . . requirements to maintain accuracy and fitness for use)

Chapter 5.9.3, *Identification,* states, "For reference purposes, a system should be established for identifying calibration procedures."

Finally, M1-1996, Chapter 4.9, states, "Documented procedures, of sufficient detail to ensure that calibrations are performed with repeatable accuracy, shall be utilized for the calibration of all ensembles." That explains it all in one sentence. (This standard describes a calibration system as an *ensemble*.)

Now that you know what the standards require for calibration procedures, where do you go from here? Some companies use original equipment manufacturer (OEM) procedures that are in their service manuals as a starting point. Please keep in mind that a lot of the service manuals have procedures for adjusting the IM&TE (inspection, measuring, and test equipment) as well as (or instead of) the calibration (performance verification) process. Also, some OEM procedures are vague and lack specific requirements needed to ensure a good calibration, such as equipment requirements, environmental conditions, and so on. Finally, in many cases the equipment manufacturer simply does not provide any calibration or service information as a matter of policy. By writing your own procedures or using pre-written calibration procedures, you might save time by eliminating the adjustment process if it is not required and/or improves the outcome of the calibration.

With regard to adjusting IM&TE, there are several schools of thought on the issue. On one end of the spectrum, some (particularly government regulatory agencies) require that an instrument be adjusted at every calibration, whether or not it is actually required. At the other end of the spectrum, some hold that any adjustment is tampering with the natural system (from Deming) and what should be done is simply to record the values and make corrections to measurements.

An intermediate position is to adjust the instrument only if (a) the measurement is outside the specification limits, (b) the measurement is inside but near the specification limits, where *near* is defined by the uncertainty of the calibration standards, or (c) a documented history of the values of the measured parameter shows that the measurement trend is likely to take it out of specification before the next calibration due date.[1]

The following two pages are an example of a calibration procedure, showing what is required, and one way to format the procedure.

Title:		Procedure No.	Rev. No.
Balance and Scale Calibration Procedure		SOP11C002	02
Submitted by: Willem Lars Public	Date: 7/10/2006	Approved by: Ayumi Jane Deaux	

READ THE ENTIRE PROCEDURE BEFORE BEGINNING

1. PURPOSE

This Standard Operating Procedure (SOP) describes the responsibilities of the Calibration Department as they relate to the calibration of all balances and scales. The intent of this SOP is to give the reader an idea of how to format and structure a calibration procedure.

2. SCOPE

This SOP applies to all balances and scales that impact the quality of goods supplied by the Mad City Widget Company, Eat-More-Cheese, Wisconsin.

3. RESPONSIBILITIES

3.1 It is the responsibility of all calibration technicians who calibrate balances and scales to comply with this SOP.

3.2 The person responsible for the repair and calibration of the balance or scale will wear rubber gloves and eye protection. The balance or scale must be cleaned and/or decontaminated by the user or owner before work can be accomplished.

4. DEFINITIONS

4.1 NIST—National Institute of Standards and Technology

4.2 TEST EQUIPMENT—Inspection, Measurement, and Test Instrument

TEST EQUIPMENT SPECIFICATIONS

Manuf.	P/N	Usable Range	Accuracy
Allied	7206A	500 mg ~ 500 g	± 30 mg (500 mg ~ 30 g) > 30 g ± 0.1% of Rdg
Denver	400	500 mg ~ 400 g	± 20 mg (500 mg ~ 20 g) > 20 g ± 0.1% of Rdg
Ohaus	V02130	50 mg ~ 210 g	± 3 mg (50 mg ~ 3 g) > 3 g ± 0.1% of Rdg

EQUIPMENT REQUIREMENTS (Calibration Standards)

Weight Size	Accuracy (mg)	Class
25 kilograms	± 2.5 grams	F
5 kilograms	± 12.0	1
50 milligrams	± 0.01	1
1 milligrams	± 0.01	1

5. PROCEDURE

    5.1. General inspection

        5.1.1. This is to give the reader an idea of a numbering system and a formatting scheme to use for calibration procedures.

    5.2. Leveling the test instrument

        5.2.1. Be as specific as possible in your instructions. Write for the benefit of the least experienced technician, not the senior person on your staff.

    5.3. Calibrating the edges of the weighing pan

        5.3.1. Following the example illustrated below, place a weight equal to approximately one half the capacity of the test instrument on the top edge of the weighing pan (place the single weight half the distance between the pan center and the usable edge). Record the reading on the calibration worksheet. Repeat for the three other edges.

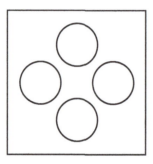 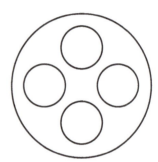

    5.4. Completing the calibration worksheet or form

        5.4.1. Identify the standard weight(s) used for each weight check by the identification number(s) in the block provided.

6. RELATED PROCEDURES

7. FORMS AND RECORDS

8. DOCUMENT HISTORY

Rev. #	Change Summary
00	New document
01	Added edge calibration verbiage.1
02	Changed accuracies to be ±0.1% of reading and instituted usable range at low and high end of all balances.

DISCONNECT AND SECURE ALL IM&TE

So what does the future hold for calibration procedures? They will never go away. There can be no doubt about that. However, having them in their current form will probably change, even in the near future.

Electronic copies have been around for some time, but the ability to read them from a computer screen may not be as easy or convenient as using a hard copy in a binder or folder. Having an automated program run in the background of a computer has also been around for quite a while, but there still has to be some type of instruction used for setup, tear down, and validation.

It is not hard to visualize wearing some type of glasses or stand-alone heads-up display (HUD) that scrolls the calibration procedure in front of you. HUDs are currently used in cars, aircraft, and so on. Why not use them to automate, increase efficiency, and reduce paper for the calibration process? The main unit housing automated or electronic calibration procedures could be programmed to use only the latest authorized version of a procedure, eliminating the need to check a roster or document control chart to ensure the calibration technician is using the current edition. Also, by going paperless, the requirement of having to file hard copies used in multiple locations is also eliminated. There are so many advantages to going paperless that it is hard to list them all.

One can also see the future of having devices that project the calibration procedure into the space directly above the item to be calibrated. This projection could come from any device that is portable and runs on batteries. It could be a wireless device that accepts only the latest version of a calibration procedure, ensuring the calibration technician is up to date with the procedure used. It would save carrying a hard copy, eliminate reading from a computer screen, and allow multiple users of the same procedure at the same time.

Only our imaginations, or lack thereof, hold us back. "Necessity is the mother of invention" is as applicable today as when it was first uttered. The need is definitely there.

NOTES

1. Jay L. Bucher, *The Metrology Handbook* (Milwaukee: ASQ Quality Press, 2004), 41, 42.
2. Bucher, *The Metrology Handbook,* 43.
3. Bucher, *The Metrology Handbook,* 44.

5

Calibration Records

One might think of record keeping as a three-pronged approach: the calibration record (electronic or hard copy), the calibration label (which has the date calibrated, next due date, name of calibration technician, and a unique identification number of the test equipment), and the calibration software record (let's call it CAMS—calibration automated management system). Each is required in a quality calibration system and performs a critical function by itself.

Here is a short list of where you can find specific requirements for documenting your calibrations through the use of records.[1]

1. ANSI/ISO/IEC 17025:2005(E), paragraph 4.13.2.1, states, "The laboratory shall retain records of original observations, derived data and sufficient information to establish an audit. Trail, calibration records, staff records and a copy of each test report or calibration certificate issued, for a defined period."[2]
2. ANSI/ISO/ASQ Q10012:2003, Chapter 6.2.3, states, "Records containing information required for the operation of the measurement management system shall be maintained. Documented procedures shall ensure the identification, storage, protection, retrieval, retention time and disposition of records."
3. NCSLI RP-6, Chapter 5.6, *Records,* states, "Records should be maintained for all measurement and test equipment that is included in the calibration system."
4. M1-1996, Chapter 4.7, *Records,* states, "Records shall include, but not be limited to . . . (a) Description of equipment and unique identification; (b) Date most recent calibration was performed; (c) Indication of procedure used; (d) Calibration interval; (e) Calibration results obtained (i.e., in or out of tolerance); (f) By whom the ensemble was calibrated; and (g) Standards used. Specific record requirements for other standards or regulations can be found in Chapter 2.11, Industry specific requirements."
5. 21 CFR Sec. 820.72, Inspection, measuring, and test equipment.
   (b) Calibration. Calibration procedures shall include specific directions and limits for accuracy and precision.
      (1) Calibration standards. Calibration standards used for inspection, measuring, and test equipment shall be traceable to national or international standards.
      (2) Calibration records. The equipment identification, calibration dates, the individual performing each calibration, and the next calibration date shall be documented.

By following the guidance in Section 5.6, *Records,* of the National Conference of Standards Laboratories (NCSL) International's RP-6, Calibration Control Systems for the Biomedical and Pharmaceutical Industry, you can decide what is needed for your particular business requirements. There are four basic areas that need to be covered on every form: identification, location, calibration history, and traceability documentation. Here is a brief explanation of how a company could fill these requirements.

Identification: Use a five-digit system, where a unique number is assigned to every piece of test equipment that you support, using a chemical- and abrasive-resistant bar code label, attached where it can be easily read on the unit. This "Metrology ID number" matches the number used in CAMS, software for tracking, data collection, and historical record keeping.

Location: List the location of each item in CAMS by its room number and the department to which it is assigned. Since your technicians may go to each laboratory or room to calibrate and/or repair each piece of test equipment, the criticality of easily finding each item cannot be overemphasized. It has been found that manpower is wasted searching for test equipment that has been moved to a new location and notification is not given so that you can update the database. Having the location in your database also allows for easy sorting by area as the need arises, such as when searching for like items that are coming due for calibration during a specific time period. This equates to working smarter, not longer.

Calibration history: Record on each form when the unit was last calibrated, the new calibration date, and when it will next be due for calibration. Show what the working or reference standard read, as well as the *As Found* and, when needed, *As Left* readings of the test instrument. The standards used are identified by their ID number, range, and tolerances, and when they are next due for calibration. Use check boxes to indicate if the test instrument passed or failed its calibration. Identify the calibration procedure used for that particular calibration. And, finally, have a place for the calibration technician to sign and date the form.

Traceability documentation: On every form, have a traceability statement identifying your unbroken chain of comparisons back to NIST. Strictly adhere to using a 4:1 ratio of uncertainty and so state in your quality system and procedures (if that is the route your calibration system is using for traceability).

In order for a record to be valid, it must:

1. Identify which test equipment on which it is recording data.
   - Have a unique identification number assigned to it
   - Identify who owns it (when appropriate)
   - Test equipment part or manufacturer's number
   - Test equipment serial number and/or asset number if applicable
   - Location, when appropriate. If it is for internal use only, consider showing the department, group, or cost center that owns it, and the calibration interval assigned to it.
2. The environmental conditions during calibration are sometimes required and need an area on the record.
3. List the instrument's ranges and tolerances, when applicable or required.

4. Show traceability for the standards you are using to perform the calibration by identifying your standards and when they are currently due for calibration.
5. List the procedure used for that particular calibration, and the revision number.
6. List your standard's tolerances and/or traceability information.
7. Have an area for recording the standard's readings, as well as what the test equipment read (this is what is known as the *As Found* readings). It will also need an area for the *As Left* readings, those readings taken after repair, alignment, or adjustment of the unit. When required, the out-of-tolerance readings may be given in magnitude and direction in order to make an impact assessment.
8. List the next *Date Due Calibration,* if required. More information on the actual date (day, month, year, or just month and year) can be found in Chapter 10.
9. An area for comments or remarks when clarification of limits, repairs, or adjustments that occurred during the process of calibration is required.
10. The person performing the calibration is usually required to date and sign the record.
    - The date will be when all calibration functions have been completed. Some items take more than a day to calibrate. When this occurs, the final day of calibration is the calibration date and is also the date used for calculating when it is next due for calibration.
    - Some organizations have a requirement for a second set of eyes to audit, perform a quality assurance function, or just ensure all the data is present. Whatever the case may be, there should be a place for their signature and possibly the date that audit was completed.

According to ANSI/ISO/IEC 17025:2005(E), paragraphs 5.10.2 and 5.10.4,[2]

Each certificate of calibration shall include:

1. A title
2. Name and address of the laboratory
3. Unique identification of the certificate
4. Name and address of the customer
5. Identification of the method used
6. A description and condition of the item calibrated
7. The date(s) of calibration
8. The calibration results and units of measurement
9. The name, function, and signature of the person authorizing the certificate
10. The environmental conditions during calibration
11. The uncertainty of measurement
12. Evidence that the measurements are traceable

What are the pros and cons of the electronic collection, handling, storage, archiving, and retrieval of your records and documentation in a metrology department or calibration laboratory? The first item that comes to mind is the saving of space; that is, filing cabinets, drawers in desks, maybe a vault in the back room. Humans have been storing records in more places than can be remembered, using a large variety of devices and cabinets, since first painting thoughts on the walls of caves. Another savings is money in time spent filing; in purchasing and maintaining all the cabinets or storage devices; in purchasing all the consumables, including paper, ink cartridges, staples and staplers, file folders for seg-regating the documents, and pens to write with, just to name a few items. How valuable is your time? It takes time to manually fill out a form, compared to doing it either elec-tronically, online, or using an automated form. It takes time to correctly file hard copy documents once they have been accepted. It is also time-consuming to retrieve and refile the forms during and after an audit. In some cases, a reduction in manpower can be achieved by going electronic in data collection and storage. From a management point of view, this could mean the difference of staying in business or closing your doors.

Choose a closed-loop computer system, in which the people responsible for the con-tent of electronic records in the system control system access. This allows security to be built into the system and requires a user name and a password to gain access to the sys-tem. No one else can access the system, files, or records, as opposed to an open loop sys-tem, where anyone turning on the computer has full access.

There are many ways to set up an electronic filing system. One way is to set up a three-folder system for maintaining and using forms. The first is an electronic folder where cal-ibration record templates can be stored. It should be password-protected and accessible only by department or calibration personnel. Within that folder is a subsection where pre-filled forms can be stored. See Figure 5.1.

The next electronic folder is for completed forms that require cosigning. This folder could be called *2-B-Cosigned* and is used by the calibration technician after they have fin-ished their calibrations and completed the calibration forms. The calibration technician copies the electronic calibration form to the *2-B-Cosigned* folder via computer or a wire-less system.

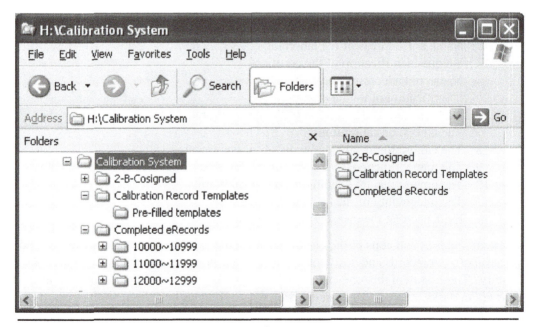

Figure 5.1 Sample calibration system computer folder.

The third folder is where all the electronic calibration records are archived after being cosigned. A numbering or archiving system should be used that allows easy access to the electronic records. A numbering format used successfully is the test equipment's identification number, followed by the Julian date that the calibration or repair work was completed. Here is an example: 13169-06031. This translates into an equipment ID number of 13169, with a calibration date of January 31, 2006. The first two digits of the Julian date are the year, with the last three the actual Julian date (the 31st day of 2006, or January 31). Using the ID number of the test equipment (it should also be the unique identifier used in CAMS), you can find the specific item being viewed. By using the Julian date protocol, you can easily identify the date needed. If the database is sorted alphabetically, it also simplifies the selection of all records for a specific piece of test equipment when required.

There are many ways an organization can go to be paperless. The various software packages available all perform basically the same function: electronic records with form fields to be completed by the calibration technician. I do not endorse or condone any particular manufacturer or product. The following example is from Adobe[3] Acrobat and happens to fill the requirements for form security and use. Acrobat Professional is required to develop a form. By using forms that have already been created in another format, such as Word, it is easy to get started. Once the form has been imported into PDF format, Acrobat is used to create form fields. Simply print your form into a PDF document. Then the PDF record can be opened using Acrobat and the process of creating a form can start. The creator decides on length, text style, size, and how many characters can be used in any particular form field. Once all of the fields are established, the creator sets up signature fields.

One of the advantages of Adobe Acrobat Pro is that the signature field may be customized to lock all of the previous fields on the form once the calibration technician's signature is used. Also, if a second set of eyes is required—someone required to cosign the form—then the entire form can also be locked once the cosigner signs. This eliminates any problem of form modification by anyone that might have electronic access to the form. Also, the form can be secured to allow only printing, viewing, and so on. This alleviates the ability of a hacker, using another copy of Acrobat Pro, to open and save the document for other purposes.

It should be noted that in order to fill in a form created with Acrobat Pro, the calibration technician must be using either Acrobat Pro or Acrobat Standard. Just using the reader version will not allow the person filling in the form to save it and its data. All they will be able to do is print the record. This is a consideration when it comes time to select software.

One definition for *electronic signature* is a computer data compilation of any symbol or series of symbols executed, adopted, or authorized by an individual to be the legally binding equivalent of the individual's handwritten signature. Many companies do not go paperless for fear of the cost of electronic signatures. If your concerns stem from GMP requirements, please consider that FDA guidelines on electronic signatures apply only if you are sending your records to FDA electronically. The requirements for 21 CFR Part 11 have been in flux for years and may have been drastically changed by the time this book goes to press. One thing is for certain: They require that your documents be stored and saved in such a manner that if they are accessed illegally, you would know and be able to take action. That much has not changed. Also, they realize that no one can have a bulletproof system, so they are not asking for that. Their requirement is for a system that can detect when changes have been made and allow the user to know this.

Most quality systems require some type of validation of records if they are maintained electronically. One way is to validate your forms using a matrix form with the following headings: Test No., Parameters, Expected Results, Actual Results, and Pass or Fail. In the Parameters block, state your objectives for that particular test. In the Expected Results block, specify what should or should not happen with that particular test. In the Actual Results block, list what happened or did not happen during the test. In the Pass/Fail block, state if that particular test passed or failed. How the electronic form is designed, protected, stored, and used can easily be validated using this type of document. It is better to be specific in stating parameters and expected results than to be general or vague. An auditor will want to see thoroughly tested electronic forms and functions rather than just accomplishing paperwork for paperwork's sake. It will also give peace of mind that your records are secure and protected through vigorous testing, and assure you of having the data available when you need it.

An important consideration in completing forms is the habit of filling in every form field available in the record. There are several reasons for this. If the record is accessed by an unauthorized person, the document could have blank form fields filled in using a typewriter. The printed record could be scanned and completed using another software program. Or data could simply be written in, signed, and dated as if the record originated there.

If all fields are completed using "N/A," no comment, not required, or any such statements, this would greatly reduce the chance of someone trying to change records. By being proactive in the completion of electronic records, the problems of altered or compromised records are almost nonexistent. But this must be impressed upon the calibration technician from the start when using electronic records. Filling in each and every block or form field is critical to record security. By training all calibration technicians about why this is important—not just telling them to do it—it will become second nature to them and make the recordkeeping system as bulletproof as possible. Figures 5.2 and 5.3 are examples of electronic forms. Part of them are pre-filled with "N/A" since they would have to be filled in if not used anyway, and this could save time for each calibration technician using these forms.

Now that electronic or paperless records have been discussed, it is time to briefly talk about going wireless in data collection and transmission. There are basically three types of wireless systems: cellular, Bluetooth, and 802.1x. A wireless router or hub is required and a wireless card is needed for the laptop. In some cases, the technology is built into the laptop by the manufacturer, as might be the case with a Bluetooth system. A company would place as many routers or hubs (they would have antennas attached or protruding from them) throughout their facility to cover the areas they would like to broadcast to and receive from.

Figure 5.4 is an example of how to use routers or hubs to set up a system. As can be seen in the example, a broadcast hub sends out its signal and receives signals in basically a predetermined area. There might be drop-out spots (areas where the signal cannot be picked up or lost) located in corners of the facility. By placing directional hubs in the

Calibration Form		
Title: Balance and Scale Calibration Form	Procedure No.: 002	Revision No.: 1

ID #: _____ P/N: _____

Range/Capacity: ±

Accuracy: ± _____

Last Cal.: _____ Today's Date: _____ Date Due Cal.: _____

Room No.: _____ User Depart. No.: _____

Interval: ☐ 1 month ☐ 6 months ☐ 12 months ☐ 18 months ☐ 24 months

Edge Test	Std Weight	TI "As Found"	TI "As Left"
1 (top)			N/A
2 (right side)			N/A
3 (bottom)			N/A
4 (left side)			N/A
ID No.	**Std Weight**	**TI "As Found"**	**TI "As Left"**
			N/A
			N/A
			N/A
			N/A
			N/A
			N/A

This test instrument was calibrated against standard weights that are traceable to NIST. *(See the reverse side of this sheet for the accuracy and calibration due dates of each standard used.)* TUR: ≥ 4:1

This TI was calibrated in accordance with SOP002.

This TI falls within specifications: ☐ Yes

☐ No (The owner/user has been notified)

Comments: _____

Calibration Technician: _____ Date: _____

Approved By: _____ Jay L. Bucher, ASQ CCT

Figure 5.2 Sample calibration form.

Calibration Form		
Title: Balance and Scale Calibration Form	Procedure No.: 002	Revision No.: 1

STANDARD WEIGHTS AND
THEIR CALIBRATION DUE DATES

ID No.	Weight Size	Accuracy (mg)	Class	Date Due Calibration
10191	25 Kg (1)	± 2.5 grams	F	
10190	25 Kg (2)	± 2.5 grams	F	
10189	25 Kg (3)	± 2.5 grams	F	
10188	25 Kg (4)	± 2.5 grams	F	
10184	5 Kg	± 12.0	1	
10183	2 Kg	± 5.0	1	
10178	1 Kg	± 2.5	1	
10176	500 grams	± 1.20	1	
10175	400 grams	± 1.00	1	
10172	300 grams	± 0.75	1	
10170	200 grams	± 0.50	1	
10169	100 grams	± 0.25	1	
10169	50 grams	± 0.120	1	
10169	20 grams	± 0.074	1	
10169	20 grams	± 0.074	1	
10169	10 grams	± 0.05	1	
10169	5 grams	± 0.034	1	
10169	2 grams	± 0.034	1	
10169	2 grams	± 0.034	1	
10169	1 gram	± 0.034	1	
10169	1~500 m/gram	± 0.01	1	

Figure 5.3 Sample calibration form.

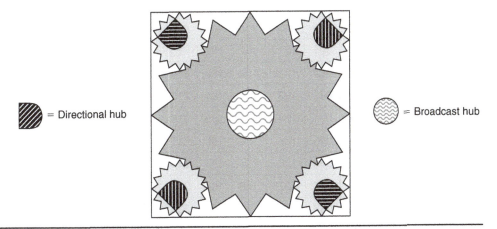

Figure 5.4 Using routers or hubs to set up a system.

drop-out areas, complete coverage is possible. Remember that it is illegal to block or interfere with any transmissions. Therefore, directional hubs become more important for those that are concerned with their data or information being broadcast by way of a wireless transmission.

My advice to the reader is to not put off going paperless another day. I can't imagine doing a job under the old system of creating hard copy records, correcting mistakes, dating and initialing those corrections, and filing all that paperwork, the vast majority of which will never be looked at again. The time saved alone is worth the effort. The money saved in paper and filing space, coupled with the advantages of increased readability of records, cannot be overemphasized.

Filling out your most frequently used forms ahead of time will also pay big dividends, in both time and accuracy. By placing repeatable information into the form (specifications, ranges and tolerances, test uncertainty ratio [TUR] or uncertainty budgets, and so on), rather than in the form field, the form is also easily updated when changes or improvements are needed on the form. Items such as balances, scales, and power supplies are examples of items that have identical requirements in a general perspective, but small differences in the tolerances. By pre-filling out the forms with each unit's tolerances, time can be saved over the long run in completing the forms during calibration. Other items like pressure and vacuum gages, voltmeters, ohmmeters, and other pieces of test equipment that are grouped into specific categories can be pre-filled, saving time and effort over the course of repeated calibrations.

The more you can incorporate your electronic records into an automated system, the more time and money you'll save. Integration of calibration software into CAMS allows shortcuts that have proven to significantly save time and money.

During audits, the ease of showing an unbroken chain of comparisons is quick and easy. As long as all the required data and information is incorporated into your electronic form template, all requirements will be met and there will be no problems complying with any regulations or system requirements.

Will there always be a requirement for records? It's hard to imagine a situation where records would not have a place in any type of quality calibration system. Electronic

Day	Jan	Feb	Mar	Apr	May	Jun	Jul	Aug	Sep	Oct	Nov	Dec	Day
1	001	032	060	091	121	152	182	213	244	274	305	335	1
2	002	033	061	092	122	153	183	214	245	275	306	336	2
3	003	034	062	093	123	154	184	215	246	276	307	337	3
4	004	035	063	094	124	155	185	216	247	277	308	338	4
5	005	036	064	095	125	156	186	217	248	278	309	339	5
6	006	037	065	096	126	157	187	218	249	279	310	340	6
7	007	038	066	097	127	158	188	219	250	280	311	341	7
8	008	039	067	098	128	159	189	220	251	281	312	342	8
9	009	040	068	099	129	160	190	221	252	282	313	343	9
10	010	041	069	100	130	161	191	222	253	283	314	344	10
11	011	042	070	101	131	162	192	223	254	284	315	345	11
12	012	043	071	102	132	163	193	224	255	285	316	346	12
13	013	044	072	103	133	164	194	225	256	286	317	347	13
14	014	045	073	104	134	165	195	226	257	287	318	348	14
15	015	046	074	105	135	166	196	227	258	288	319	349	15
16	016	047	075	106	136	167	197	228	259	289	320	350	16
17	017	048	076	107	137	168	198	229	260	290	321	351	17
18	018	049	077	108	138	169	199	230	261	291	322	352	18
19	019	050	078	109	139	170	200	231	262	292	323	353	19
20	020	051	079	110	140	171	201	232	263	293	324	354	20
21	021	052	080	111	141	172	202	233	264	294	235	255	21
22	022	053	081	112	142	173	203	234	265	295	326	356	22
23	023	054	082	113	143	174	204	235	266	296	327	357	23
24	024	055	083	114	144	175	205	236	267	297	328	358	24
25	025	056	084	115	145	176	206	237	268	298	329	359	25
26	026	057	085	116	146	177	207	238	269	299	330	360	26
27	027	058	086	117	147	178	208	239	270	300	331	361	27
28	028	059	087	118	148	179	209	240	271	301	332	362	28
29	029		088	119	149	180	210	241	272	302	333	363	29
30	030		089	120	150	181	211	242	273	303	334	364	30
31	031		090		151		212	243		304		365	31

Figure 5.5   Julian date calendar.

Day	Jan	Feb	Mar	Apr	May	Jun	Jul	Aug	Sep	Oct	Nov	Dec	Day
1	001	032	061	092	122	153	183	214	245	275	306	336	1
2	002	033	062	093	123	154	184	215	246	276	307	337	2
3	003	034	063	094	124	155	185	216	247	277	308	338	3
4	004	035	064	095	125	156	186	217	248	278	309	339	4
5	005	036	065	096	126	157	187	218	249	279	310	340	5
6	006	037	066	097	127	158	188	219	250	280	311	341	6
7	007	038	067	098	128	159	189	220	251	281	312	342	7
8	008	039	068	099	129	160	190	221	252	282	313	343	8
9	009	040	069	100	130	161	191	222	253	283	314	344	9
10	010	041	070	101	131	162	192	223	254	284	315	345	10
11	011	042	071	102	132	163	193	224	255	285	316	346	11
12	012	043	072	103	133	164	194	225	256	286	317	347	12
13	013	044	073	104	134	165	195	226	257	287	318	348	13
14	014	045	074	105	135	166	196	227	258	288	319	349	14
15	015	046	075	106	136	167	197	228	259	289	320	350	15
16	016	047	076	107	137	168	198	229	260	290	321	351	16
17	017	048	077	108	138	169	199	230	261	291	322	352	17
18	018	049	078	109	139	170	200	231	262	292	323	353	18
19	019	050	079	110	140	171	201	232	263	293	324	354	19
20	020	051	080	111	141	172	202	233	264	294	325	355	20
21	021	052	081	112	142	173	203	234	265	295	326	356	21
22	022	053	082	113	143	174	204	235	266	296	327	357	22
23	023	054	083	114	144	175	205	236	267	297	328	358	23
24	024	055	084	115	145	176	206	237	268	298	329	359	24
25	025	056	085	116	146	177	207	238	269	299	330	360	25
26	026	057	086	117	147	178	208	239	270	300	331	361	26
27	027	058	087	118	148	179	209	240	271	301	332	362	27
28	028	059	088	119	149	180	210	241	272	302	333	363	28
29	029	060	089	120	150	181	211	242	273	303	334	364	29
30	030		090	121	151	182	212	243	274	304	335	365	30
31	031		091		152		213	244		305		366	31

Figure 5.6   Julian date calendar for leap years.

records, as well as using a wireless system, have already been discussed. How can improvements be made to either of these systems? Being completely automated and built into the firmware of the test instrument would be a start. There are already some on the market today. Questions that arise concern validation of software and/or qualification of the test equipment. This must be accomplished before either could be used in a regulated environment.

As we become more and more automated, there could come a day when the data from a calibration record is automatically sent to a database used for analysis of product or production (see the end of Chapter 6). If a product must be recalled, the data from every prior calibration could be used to show compliance to requirements or that the effect on product or processes was or was not applicable.

The more automated and computerized records and data become, the easier it will be to evaluate large blocks of data without having to pull records, copy data, and compile information. The need to have computers and automation work with us, instead of just for us, might become the driving factor in some organizations to go paperless and/or wireless in the record and data collection, storage, and archiving.

Figures 5.5 and 5.6 are of Julian date calendars. These are provided for use in setting up record file names as demonstrated earlier in this chapter. By selecting the day of the month by going down the left or right hand column, and the month of the year from across the top, the Julian date can easily be found. Figure 5.6 is used only during leap year (2000, 2004, 2008, and so on).

NOTES

1. Jay L. Bucher, *The Metrology Handbook* (Milwaukee: ASQ Quality Press, 2004), 48.

2. ANSI/ISO/IEC, *ANSI/ISO/IEC 17025-2005: General requirements for the competence of testing and calibration laboratories* (Milwaukee: ASQ Quality Press, 2005).

3. Adobe, http://www.adobe.com (22 August 2006).

6

It's Out of Tolerance—Now What?

When test equipment is found to be out of tolerance during calibration (or when it comes in for repair or adjustment, and an *As Found* calibration is performed and an out-of-tolerance condition is recorded), your quality calibration program must have a system in place for notification, recall, root cause analysis, and so on.

Within your quality system, does it make a difference how far out of tolerance the test equipment is before anyone gets notified? What documentation is involved and who is responsible for starting the process? How is test equipment recalled? Who determines if product must be recalled? A lot of questions, and for most calibration departments, not enough answers. Some of the answers can be found in this chapter.

When a calibration step does not meet the specified tolerances, the calibration must be continued until all parameters have been calibrated (compared against your standard). Make a note about the out-of-tolerance step by recording what the test instrument read and what the standard read. But the entire calibration must be completed, because if the calibration is stopped at that point and an adjustment is made, it is possible that the adjustment could affect other parts of the calibration, both for measurements that have already been made or have yet to be done. If they have not been done yet, they could show up as in-tolerance when in fact they were out of tolerance before the adjustment, and it would not be possible to inform the user of this out-of-tolerance condition. They could show as out of tolerance when they were originally within tolerance and the user has to make a false recall or do-over because of bad data.

Here are some examples from the federal regulations on requirements when test equipment is found to be out of tolerance (underscore emphasis added):

Sec. 211.160 General requirements.
    Para:(b)(4) The calibration of instruments, apparatus, gauges, and recording devices at suitable intervals in accordance with an established written program containing specific directions, schedules, limits for accuracy and precision, and provisions for remedial action in the event accuracy and/or precision limits are not met. Instruments, apparatus, gauges, and recording devices not meeting established specifications shall not be used.
Sec. 820.72 Inspection, measuring, and test equipment.
    Para:(b) Calibration. Calibration procedures shall include specific directions and limits for accuracy and precision. When accuracy and precision limits are not met,

there shall be provisions for remedial action to reestablish the limits and to evaluate whether there was any adverse effect on the device's quality. These activities shall be documented.

Sec. 58.63 Maintenance and calibration of equipment.

Para:(c) Written records shall be maintained of all inspection, maintenance, testing, calibrating, and/or standardizing operations. These records, containing the date of the operation, shall describe whether the maintenance operations were routine and followed the written standard operating procedures. Written records shall be kept of nonroutine repairs performed on equipment as a result of failure and malfunction. Such records shall document the nature of the defect, how and when the defect was discovered, and any remedial action taken in response to the defect.

As clearly written in these federal regulations, whenever something is found to be out of tolerance, it will be recorded, and any remedial action shall also be recorded. At a minimum, the *As Found* reading from the test instrument, as well as what the working or reference standard read should be recorded, along with the *As Left* reading, after the test instrument was repaired and/or adjusted.

Management determines any additional information that an organization keeps, in what format it is retained, and how the record is archived. Figure 6.1 is an example of an electronic form that could be used to inform the customer that their test equipment was found to be out of tolerance.

Figure 6.1 displays various information about the test instrument that was found to be out of tolerance, including:

- The ID number
- Part number
- Type of equipment
- Range and capacity, depending on the type of equipment
- Accuracy or tolerance that the test instrument is expected to meet
- The last time it was calibrated
- The date the test instrument was found to be out of tolerance
- Where the test instrument was located
- The customer's department or division identifier
- The calibration interval of the test instrument
- What the working or reference standard read during each step of the calibration, to include both the in-tolerance and out-of-tolerance ranges
- What the test instrument read during the *As Found* or initial calibration readings
- What the test instrument read during the *As Left* readings after repair/adjustment of the test instrument were completed

As can be seen in Figure 6.1, there are both *Alert* and *Action* areas. In some companies, it has been found during risk analysis that there is little or no risk of impact on product or processes if test equipment is out of tolerance by a certain amount. The amount in the example has been set at twice the tolerance.

What that means is that if a tolerance is ± 2 °C, then anything out of tolerance greater than ± 4 °C would require an impact assessment by the customer. Any out-of-tolerance

Test Instrument (TI) Alert/Action Procedure Form

ID #:_____   P/N: _____   Type of Equipment: _____

Range/Capacity: _____   Accuracy: ± _____

Last Cal: _____   Today's Date: _____

Room No.:_____   User Department: _____   Calibration Interval:_____ months

Data Collected From the Metrology Calibration Worksheet

Working Standard Reading	Test Instrument As Found Reading	Test Instrument As Left Reading

☐ **Alert**   (out of tolerance by ≤ 2X the TI tolerances)

☐ **Action**   (out of tolerance by > 2X the TI tolerances)

☐ This test instrument was adjusted to meet specifications

☐ This test instrument does not meet specifications and received a limited certification

☐ This test instrument is not adjustable/repairable and has been removed from service

Comments: _____

Metrologist (typed name): _____   Date Completed: _____

CAMS Updated: ☐ Yes ☐ No   Alert/Action Control Log Updated: ☐ Yes ☐ No

This form serves to notify the user/owner of test equipment found to be out of tolerance during calibration. The user/owner of this test instrument is responsible for determining if the use of this item during the out-of-tolerance period adversely affected product development or manufacturing processes, as appropriate.

The following section to be completed by the customer only

Project/Lot #: _____   Project/Product Type: _____

How was the test instrument used?_____

Product or process impact: _____

Summary/Recommendations: _____

Assessment conducted by: _____   Date: _____

This assessment has been: ☐ Approved   ☐ Disapproved

Area Supervisor/Manager/Director/VP: _____   Date: _____

Figure 6.1   Sample out-of-tolerance notification form.

ID No.	Alert	Action	Status	Type	Model	Range	Out by	Date	Room	Dept.	Contact person	Comments
69169		XXX	Closed									
69238		XXX	Open									
69307	XXX		N/A									
69376	XXX		N/A									
69445		XXX	Closed									
69514	XXX		N/A									
69583	XXX		N/A									
69652	XXX		N/A									
69721		XXX	Closed									

Figure 6.2 Out-of-tolerance tracking database.

condition of less than ± 4 °C would only require the customer be notified of the out-of-tolerance condition, but an impact assessment does not have to be made. It must be remembered that these conditions would fall within the confines of a quality calibration system used within a company for their own test equipment, not for third-party calibration laboratories doing commercial work for outside customers.

A risk analysis must be performed at your company to decide what this cut-off point should be, or even if you will have a cut-off point. In some instances, any occurrence of out-of-tolerance condition would result in impact assessment of product or process. Also, after the Alert/Action check boxes one might notice the following three lines:

- This test instrument was adjusted to meet specifications.
- This test instrument does not meet specifications and received a limited certification.
- This test instrument is not adjustable/repairable and has been removed from service.

If the test instrument was out of tolerance, there are three general choices for solving the problem. The test instrument can be adjusted, aligned, and/or repaired (not necessarily in that order) and then receive another calibration before being put back into service. In the case where the test instrument possibly cannot be adjusted for various reasons:

- The range or parameter that is out of tolerance is not needed by the user and a limited calibration would save time and money by not performing any repairs.
- The accuracy that the test instrument was calibrated against initially was not required by the user, so a limited calibration is given to the item, decreasing the accuracy on this particular test instrument, while still meeting the requirements of the user.
- Parts or labor are not available to adjust or repair the test instrument, and the customer can use the test instrument in its current state without detriment to their operation.

There are times when test equipment is worn out. It becomes too expensive to procure replacement parts, parts are simply no longer available from the manufacturer or other sources, or the technology of the item is no longer supported or used and the item needs to be disposed of. In any case, the disposition of the test instrument must be documented in CAMS for historical purposes, and the instrument must be removed from service.

Figure 6.2 is an example of a database used to track or monitor test equipment found to be out of tolerance. This particular chart illustrates a system that uses an Alert/Action program. In this type of program, *Alerts* are items that are out of tolerance by a specific amount not to exceed x-amount, but the customer is still notified that the test equipment was out of tolerance. *Action* items are those that exceed the *Alerts* and would require action by the customer or user of the test equipment, who is also notified of the out-of-tolerance condition.

The various items in the chart include:

- A column for the test equipment's ID number or unique identifier
- A column to show if it is an Alert or Action item
- A column to show the status—in the case of Alert, no action is needed, but Action items need to be closed out by way of the Action Assessment documents being

returned for archiving and a responsible party updating the database to show that each Action item has been closed

- A column for the type of test equipment that was out of tolerance
- A column for model numbers or more specific identifiers of the test equipment
- A column for the range or parameter that was out of tolerance
- A column for *Out of Tolerance by* in case the exact data is required
- A column for the date the test equipment was found to be out of tolerance
- A column for the room or location where the test instrument is normally kept
- A column for the department or division designator when required
- A column for a contact person, or the person the Action Assessment was sent to
- A column for comments or remarks that could provide additional information that might not be found anywhere else

Every organization will have specific requirements for the information that should be in this particular kind of database. The determining factors should be the standard or regulations that the company is complying with, or any special information that might assist in future determinations or qualifications of their test equipment.

This information can also be used during calibration interval analysis in helping determine test equipment pass rates. More information on calibration intervals can be found in Chapter 15. Some organizations like to keep track of their Alert/Actions as a metric in their monthly reports. There are also those that post their Alert/Actions on internal Web sites to give a visual presentation of how many pieces of test equipment have been found to be out of tolerance during the previous months. Figure 6.3 is an example of a 12-month look at Alert/Actions in chart form.

A chart as in Figure 6.3 could give the viewer a quick visual representation of how many out-of-tolerance items are found each month. Of course, it doesn't do much good without knowing the total production or amount of test equipment calibrated during the same time periods. But at least it is a quick way to see if there are any negative trends in progress during specific time periods or during certain quarters. As mentioned previously, this data can be incorporated into your calibration pass rate to help determine calibration intervals.

What does the future hold for those that calibrate and are responsible for letting their customers know their test equipment is found to be out of tolerance? The possibilities are endless. The criticality of knowing when test equipment is out of tolerance, and how that may have affected a manufacturing process, produced items, or sold goods makes the most hardened CEO cringe. The cost of recalls and lost customer confidence can be incalculable in some circumstances.

The day is not far off when the discovery of out-of-tolerance test equipment will automatically trigger analysis of any item the unit was used for in manufacturing or production. Simply inputting a code against that particular piece of test equipment would list where, when, and how the item was involved in production. An automatic analysis of effect on production could be performed, with almost instant data generated showing lot and batch numbers, where the product is at the current time, and if recall is needed.

With the increasing demand for automation and computerization of production and manufacturing, we are only one or two steps from being able to have this type of process

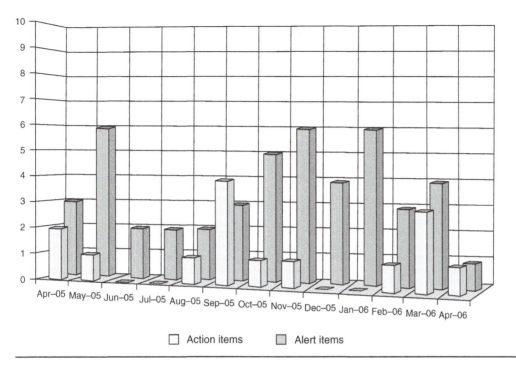

Figure 6.3 12-month alerts.

in place. Marrying the production process to a calibration department's CAMS would be simple enough. The tracking of test equipment used during R&D, manufacturing, shipping, and storage is already a requirement in some industries.

By connecting all of these processes into one database, it would be simple to generate a recall list of all the processes that involved a particular piece of test equipment and associated production lots and batches to initiate some type of analysis of that product. There may be systems already in place that are being kept from public view because they could provide a major innovative advantage over the competition. We can only dream. In some cases, the dreams may have already come true.

7

Calibration Standards

Aharmonized measurement unit ensures that everybody concerned with a measurement result understands it the same way. For acceptability, it is also essential that the measurement results for the same parameter measured at different places and by different people be in agreement. This implies that measurement results should be correlated. To achieve this agreement in measurement results, it is essential that everybody draw their measurement units from a common acceptable standard. Thus, a *standard* is a physical object or a characteristic of a physical apparatus that represents the conceptual unit chosen to represent a particular measurable attribute. As mentioned previously, measurement is critical to trade, industry, and government worldwide. The same can be said for measurement standards.

It was a common belief among calibration technicians in U.S. Air Force PMELs (Precision Measurement Equipment Laboratories) that test equipment calibrated at one PMEL would get the same results if calibrated at another PMEL, no matter if it was on the same base or halfway around the world. The same is true in most industries: an item calibrated at one location should be able to make a like measurement anywhere within that system. This is possible because measurement systems that use standards traceable to a national or international standard will form the foundation for all measurements used within that SI (international system of weights and measures). As one can see, calibration standards are the basis for all of our measurements, no matter how large or small the system, measurement, or equipment.

A *unit of measurement* has been defined as a particular quantity defined and adopted by convention, with which other quantities of the same kind are compared in order to express their magnitude relative to that quantity. The SI has been defined as the coherent system of units adopted and recommended by the GMCP (General Conference on Weights and Measures). SI units consist of seven base units and two supplementary units. The base units are regarded as dimensionally independent, and all other units are derived from the base units or from other derived units. The base units consist of:[1]

length (the meter—m)
The 1889 definition of the metre, based on the international prototype of platinum-iridium, was replaced by the 11th CGPM (1960) using a definition based on the wavelength of krypton 86 radiation. This change was adopted in order to improve the accuracy with which the definition of the metre could be realized, the realization being achieved using an interferometer with a traveling microscope to measure the optical

path difference as the fringes were counted. In turn, this was replaced in 1983 by the 17th CGPM (1983, Resolution 1), which specified the current definition as "the length of the path traveled by light in vacuum during a time interval of 1/299,792,458 of a second." It follows that the speed of light in vacuum is exactly 299,792,458 metres per second, $c_0 = 299\ 792\ 458$ m/s.

The original international prototype of the metre, which was sanctioned by the 1st CGPM in 1889, is still kept at the BIPM under conditions specified in 1889.

mass (kilogram—kg)

The international prototype of the kilogram, an artifact made of platinum-iridium, is kept at the BIPM (International Bureau of Weights and Measures) under the conditions specified by the 1st CGPM in 1889 when it sanctioned the prototype and declared:

> This prototype shall henceforth be considered to be the unit of mass.

The 3rd CGPM (1901), in a declaration intended to end the ambiguity in popular usage concerning the use of the word *weight,* confirmed that "the kilogram is the unit of mass; it is equal to the mass of the international prototype of the kilogram."

It follows that the mass of the international prototype of the kilogram is always 1 kilogram exactly, $m(K) = 1$ kg. However, due to the inevitable accumulation of contaminants on surfaces, the international prototype is subject to reversible surface contamination that approaches 1 μg per year in mass. For this reason, the International Committee of Weights and Measures "CIPM" declared that, pending further research, the reference mass of the international prototype is that immediately after cleaning and washing by a specified method (PV, 1989, 57, 104–105 and PV, 1990, 58, 95–97). The reference mass thus defined is used to calibrate national standards of platinum-iridium alloy (*Metrologia,* 1994, 31, 317–336).

time (second—s)

The unit of time, the second, was at one time considered to be the fraction 1/86,400 of the mean solar day. The exact definition of *mean solar day* was left to the astronomers. However, measurements showed that irregularities in the rotation of the Earth made this an unsatisfactory definition. In order to define the unit of time more precisely, the 11th CGPM (1960, Resolution 9) adopted a definition given by the International Astronomical Union based on the tropical year 1900. Experimental work, however, had already shown that an atomic standard of time, based on a transition between two energy levels of an atom or a molecule, could be realized and reproduced much more accurately. Considering that a very precise definition of the unit of time is indispensable for science and technology, the 13th CGPM (1967/68, Resolution 1) replaced the definition of the second with the following:

> The second is the duration of 9,192,631,770 periods of the radiation corresponding to the transition between the two hyperfine levels of the ground state of the caesium 133 atom.

It follows that the hyperfine splitting in the ground state of the caesium 133 atom is exactly 9,192,631,770 hertz, $\nu(\text{hfs Cs}) = 9{,}192{,}631{,}770$ Hz.

At its 1997 meeting, the CIPM affirmed, "this definition refers to a caesium atom at rest at a temperature of 0 K."

This note was intended to make it clear that the definition of the SI second is based on a caesium atom unperturbed by black body radiation, that is, in an environment whose thermodynamic temperature is 0 K. The frequencies of all primary frequency standards should therefore be corrected for the shift due to ambient radiation, as stated at the meeting of the Consultative Committee for Time and Frequency in 1999.

electric current (**ampere—A**)

Electric units, called *international units,* for current and resistance were introduced by the International Electrical Congress held in Chicago in 1893, and definitions of the *international ampere* and *international ohm* were confirmed by the International Conference in London in 1908.

Although it was already obvious on the occasion of the 8th CGPM (1933) that there was a unanimous desire to replace those international units by so-called *absolute units,* the official decision to abolish them was only taken by the 9th CGPM (1948), which adopted the ampere for the unit of electric current, following a definition proposed by the CIPM (1946, Resolution 2):

> The ampere is that constant current which, if maintained in two straight parallel conductors of infinite length, of negligible circular cross-section, and placed 1 metre apart in vacuum, would produce between these conductors a force equal to 2×10^{-7} newton per metre of length.

It follows that the magnetic constant, μ_0, also known as the permeability of free space, is exactly $4\pi \times 10^{-7}$ henries per metre, $\mu_0 = 4\pi \times 10^{-7}$ H/m. The expression *MKS unit of force,* which occurs in the original text of 1946, has been replaced here by *newton,* a name adopted for this unit by the 9th CGPM (1948, Resolution 7).

thermodynamic temperature (**Kelvin—K**)

The definition of the unit of thermodynamic temperature was given in substance by the 10th CGPM (1954, Resolution 3), which selected the triple point of water as the fundamental fixed point and assigned to it the temperature 273.16 K, so defining the unit. The 13th CGPM (1967/68, Resolution 3) adopted the name kelvin, symbol K, instead of *degree Kelvin,* symbol °K, and defined the unit of thermodynamic temperature as follows (1967/68, Resolution 4): "The kelvin, unit of thermodynamic temperature, is the fraction 1/273.16 of the thermodynamic temperature of the triple point of water."

It follows that the thermodynamic temperature of the triple point of water is exactly 273.16 kelvins, $T_{\text{tpw}} = 273.16$ K.

At its 2005 meeting the CIPM affirmed that:

> This definition refers to water having the isotopic composition defined exactly by the following amount of substance ratios: 0.000 155 76 mole of ^{2}H per mole of ^{1}H, 0.000 379 9 mole of ^{17}O per mole of ^{16}O, and 0.002 005 2 mole of ^{18}O per mole of ^{16}O.

Because of the manner in which temperature scales were previously defined, it remains common practice to express a thermodynamic temperature, symbol *T,* in terms of its difference from the reference temperature $T_0 = 273.15$ K, the ice point. This difference is called the Celsius temperature, symbol t, which is defined by the quantity equation:

$$t = T - T_0.$$

The unit of Celsius temperature is the degree Celsius, symbol °C, which is by definition equal in magnitude to the kelvin. A difference or interval of temperature may be expressed in kelvins or in degrees Celsius (13th CGPM, 1967/68, Resolution 3, mentioned above), the numerical value of the temperature difference being the same. However, the numerical value of a Celsius temperature expressed in degrees Celsius is related to the numerical value of the thermodynamic temperature expressed in kelvins by the relation:

$$t/°C = T/K - 273.15.$$

The kelvin and the degree Celsius are also units of the International Temperature Scale of 1990 (ITS-90) adopted by the CIPM in 1989 in its Recommendation 5 (CI-1989).

luminous intensity (candela—cd)

The units of luminous intensity based on flame or incandescent filament standards in use in various countries before 1948 were replaced initially by the *new candle* based on the luminance of a Planck radiator (a black body) at the temperature of freezing platinum. This modification had been prepared by the International Commission on Illumination (CIE) and by the CIPM before 1937, and the decision was promulgated by the CIPM in 1946. It was then ratified in 1948 by the 9th CGPM, which adopted a new international name for this unit, the *candela,* symbol cd; in 1967 the 13th CGPM (Resolution 5) gave an amended version of this definition.

In 1979, because of the difficulties in realizing a Planck radiator at high temperatures, and the new possibilities offered by radiometry, that is, the measurement of optical radiation power, the 16th CGPM (1979, Resolution 3) adopted a new definition of the candela:

> The candela is the luminous intensity, in a given direction, of a source that emits monochromatic radiation of frequency 540 (10^{12} hertz and that has a radiant intensity in that direction of 1/683 watt per steradian.

It follows that the spectral luminous efficacy for monochromatic radiation of frequency of 540×10^{12} hertz is exactly 683 lumens per watt, $K = 683$ lm/W = 683 cd sr/W.

amount of a substance (mole—mol)

Following the discovery of the fundamental laws of chemistry, units called, for example, *gram-atom* and *gram-molecule,* were used to specify amounts of chemical elements or compounds. These units had a direct connection with *atomic weights* and *molecular weights,* which are in fact relative masses. Atomic weights were originally referred to as the atomic weight of oxygen, by general agreement taken as 16. But whereas physicists separated the isotopes in a mass spectrometer and attributed the value 16 to one of the isotopes of oxygen, chemists attributed the same value to the (slightly variable) mixture of isotopes 16, 17, and 18, which was for them the naturally occurring element oxygen. Finally an agreement between the International Union of Pure and Applied Physics (IUPAP) and the International Union of Pure and Applied Chemistry (IUPAC) brought this duality to an end in 1959–1960. Physicists and chemists have ever since agreed to assign the value 12, exactly, to the so-called atomic weight of the isotope of carbon with mass number 12 (carbon 12, ^{12}C), correctly called the relative atomic mass $A_r(^{12}C)$. The unified scale thus obtained gives the relative atomic and molecular masses, also known as the atomic and molecular weights, respectively.

The quantity used by chemists to specify the amount of chemical elements or compounds is now called *amount of substance*. Amount of substance is defined to be proportional to the number of specified elementary entities in a sample, the proportionality constant being a universal constant which is the same for all samples. The unit of amount of substance is called the *mole,* symbol mol, and the mole is defined by specifying the mass of carbon 12 that constitutes one mole of carbon 12 atoms. By international agreement this was fixed at 0.012 kg, that is, 12 g.

Following proposals by the IUPAP, the IUPAC, and the ISO, the CIPM gave a definition of the mole in 1967 and confirmed it in 1969. This was adopted by the 14th CGPM (1971, Resolution 3):

1. The mole is the amount of substance of a system which contains as many elementary entities as there are atoms in 0.012 kilogram of carbon 12; its symbol is *mol.*
2. When the mole is used, the elementary entities must be specified and may be atoms, molecules, ions, electrons, other particles, or specified groups of such particles.

It follows that the molar mass of carbon 12 is exactly 12 grams per mole, $M(^{12}C) = 12$ g/mol.

In 1980 the CIPM approved the report of the CCU (1980) that specified, "in this definition, it is understood that unbound atoms of carbon 12, at rest and in their ground state, are referred to."

The definition of the mole also determines the value of the universal constant that relates the number of entities to amount of substance for any sample. This constant is called the Avogadro constant, symbol N_A or L. If $N(X)$ denotes the number of entities X in a specified sample, and if $n(X)$ denotes the amount of substance of entities X in the same sample, the relation is

$$n(X) = N(X)/N_A.$$

Note that since $N(X)$ is dimensionless, and $n(X)$ has the SI unit mole, the Avogadro constant has the coherent SI unit reciprocal mole.

In the name *amount of substance*, the words *of substance* could for simplicity be replaced by words to specify the substance concerned in any particular application, so that one may, for example, talk of amount of hydrogen chloride, HCl, or amount of benzene, C_6H_6. It is important to always give a precise specification of the entity involved (as emphasized in the second sentence of the definition of the mole); this should preferably be done by giving the empirical chemical formula of the material involved. Although the word *amount* has a more general dictionary definition, this abbreviation of the full name *amount of substance* may be used for brevity. This also applies to derived quantities such as *amount of substance concentration,* which may simply be called *amount concentration.* However, in the field of clinical chemistry the name *amount of substance concentration* is generally abbreviated to *substance concentration.*

The SI has 19 derived units that are obtained by forming various combinations of the base units, supplementary units, and other derived units. Their names, symbols, and how their values are obtained can be found in Table 7.1.[2] Table 7.2 details SI prefixes.

It is well known that SI units have been accepted internationally and are the basis of modern measurements for governments, academia, and industry. The evolution of practical national and international measurement systems is achieved in four stages.[3]

- *Definition of the unit:* The accuracy of the definition of a unit is important, as it will be reflected in the accuracy of the measurements that can be achieved.
- *Realization of the unit:* The definition of the unit has to be realized so that it can be used as a reference for measurement. This task is carried out by National Metrology Institute (NMI). The units are realized in the form of experimental setups.
- *Representation of the unit:* The realized experimental setup of the system of the unit is the physical representation of the unit. NMIs are responsible for the maintenance of this representation. They ensure that these experimental setups continue to represent the SI and are available for reference.
- *Dissemination of the unit:* The end users of measurements are trade, industry, and calibration laboratories. They do not have access to the representations to the SI

Table 7.1 SI derived units.

Parameter	Unit (symbol)	Value
Frequency	Hertz (Hz)	l/s
Force	Newton (N)	$kg \cdot m/s^2$
Pressure, stress	Pascal (Pa)	N/m^2
Energy, work, quantity of heat	Joule (J)	N(m
Power, radiant flux	Watt (W)	J/s
Electric potential difference	Volt (V)	W/A
Electric resistance	Ohm (()	V/A
Electric charge	Coulomb (C)	A(s
Electric capacitance	Farad (F)	C/V
Electric conductance	Siemens (S)	A/V
Magnetic flux	Weber (Wb)	V(S
Magnetic flux density	Tesla (T)	Wb/m^2
Inductance	Henry (H)	Wb/A
Celsius temperature	Degree (°C)	–
Luminous flux	Lumen (lm)	$cd \cdot Sr$
Illuminance	Lux (lx)	Lm/m^2
Activity (of a radionuclide)	Becquerel (Bq)	–
Absorbed dose	Gray (Gy)	J/kg
Dose equivalent	Sievert (Sv)	J/kg

Table 7.2 SI prefixes.

Factor	Prefix	Symbol	Factor	Prefix	Symbol
$10^{24} = (10^3)^8$	yotta	Y	10^{-1}	deci	d
$10^{21} = (10^3)^7$	zeta	Z	10^{-2}	centi	c
$10^{18} = (10^3)^6$	exa	E	$10^{-3} = (10^3)^{-1}$	milli	m
$10^{15} = (10^3)^5$	peta	P	$10^{-6} = (10^3)^{-2}$	micro	μ
$10^{12} = (10^3)^4$	tera	T	$10^{-9} = (10^3)^{-3}$	nano	n
$10^9 = (10^3)^3$	giga	G	$10^{-12} = (10^3)^{-4}$	pico	p
$10^6 = (10^3)^2$	mega	M	$10^{-15} = (10^3)^{-5}$	femto	f
$10^3 = (10^3)^1$	kilo	k	$10^{-18} = (10^3)^{-6}$	atto	a
10^2	hector	h	$10^{-21} = (10^3)^{-7}$	zepto	z
10^1	deka	da	$10^{-24} = (10^3)^{-8}$	yocto	y

units held by NMIs. The end users also need the values of the SI units for reference. This is accomplished through the process of dissemination, wherein the units are made available to the end users of measurement results.[3]

A harmonized measurement unit ensures that everybody concerned with a measurement result understands it the same way. For acceptability, it is also essential that the measurement results for the same parameter measured at different places and by different people be in agreement. This implies that measurement results should be correlated. To achieve this agreement in measurement results, it is essential that everybody draws their measurement units from a common acceptable standard. Thus, a *standard* is a physical object or a characteristic of a physical apparatus that represents the conceptual unit chosen to represent a particular measurable attribute.[5] As mentioned previously, measurement is critical to trade, industry, and government worldwide. The same can be said for measurement standards.

A *measurement standard* has been defined as a material measure, measuring instrument, reference material, or measuring system intended to define, realize, conserve, or reproduce a unit of one or more values of a quantity to serve as a reference. There are various categories of standards used throughout our industry, and are given in Table 7.3.[2]

One of the considerations that should be made early in any quality calibration program is the use and care of the standards, both working and reference. I do not endorse any product, but only references those that have been found to work effectively from a historical perspective. In other words, this particular brand of case has been found to protect test equipment, while generally not being cost prohibitive. Where the organization purchases such cases might be the determining factor in cost if they decide to use the same type of cases.

One method to use to help keep working and reference standards safe is Pelican brand cases,[4] which work well for three distinct areas: storage of standards when not in use, transport of standards to and from the job site or between laboratories where test equipment is calibrated, and during shipment when reference standards must be sent back

Table 7.3   Definitions of various types of standards.[6]

Type of Standard	Definition	Example
International	A standard recognized by international agreement to serve internationally as the basis for fixing the value of all other standards of the quantity concerned	The prototype of the kilogram maintained at the International Bureau of Weights and Measures (BIPM) is an international standard of mass
National	A standard recognized by an official national decision to serve in a country as the basis for fixing the value of all other standards of the quantity concerned. Generally, the national standard in a country is also a primary standard to which other standards are traceable.	National prototypes of the kilogram which are identical to the international prototype of the kilogram are maintained as national standards of mass in various NMIs
Primary	A standard that is designated or widely acknowledged as having the highest metrological quality and whose value is accepted without reference to other standards of the same quantity. National standards are generally primary standards.	The metrological quality of the Josephson-junction-based voltage standard is far superior to that of the standard cell. However, it could take quite some time to replace the standard cell as the national standard of voltage. Until then it remains the primary standard.
Secondary	A standard whose value is based upon comparisons with some primary standard. Note that a secondary standard, once its value is established, can become a primary standard for some other user.	The national standard of length consists of a stabilized laser source. High-accuracy gage blocks are used as a secondary standard of length. These standards are assigned values based on their comparison with national standards.
Reference	A standard having the highest metrological quality available at a given location from which the measurements made at that location are derived.	In the United States, state legal metrology laboratories maintain NIST-calibrated kilogram standards. These serve as reference standards for them.
Working	A measurement standard not specifically reserved as a reference standard that is intended to verify measuring equipment of lower accuracy.	Multifunction calibrators are used as working standards for the calibration of TEST EQUIPMENT to be used in the measurement of various electrical parameters.
Transfer	A standard that is the same as a reference standard except that it is used to transfer a measurement parameter from one organization to another for traceability purposes.	Standard cells are used as transfer standards for the transfer of voltage parameters from the national standard to other standards.

to the manufacturer or a third-party calibration laboratory for recalibration. They are totally watertight, crushproof, and dustproof, providing a certain amount of stress relief during all three of the previously discussed times.

Their foam lining can be made to fit any size test equipment, along with accessories and options. If for some reason the standard is scrapped, junked, or thrown away, a new foam insert can be ordered and the replacement standard can be perfectly fit into the old case.

Since any organization wants to save money where it can, the purchase of protective cases for its standards is essential to their long-term use and value. The greatest killers of test equipment are dust, dirt, and water. Even though the world of electronics, dimension equipment, and chemical standards have come a long way in a very few years . . . they can still be made inoperative by the three entities mentioned above. The protection of your standards is critical to their availability when needed to support the test equipment at your facility. Get the right cases, but more importantly, use them during storage, transport, and shipment.

Another topic that is seldom discussed but critical to any calibration program is the substitution of standards, both working and reference standards. There may come a time when a particular standard is not available for use, either because it is out for calibration, being used by another calibration technician, or down for repair. No matter the reason, any calibration technician should be trained on how to substitute standards when the reference or working standard is not available.

When substituting standards, a comparison of the tolerance and ranges involved in the calibration should take place. For this example, let's call the original standard Std A, and the standard being reviewed as a substitute Std B. The ranges called for during the use of Std A must also be present in Std B. All of the tolerances listed during the calibration during the use of Std A must be met by Std B, or exceeded.

An example would be the tolerance of Std A for the calibration of temperature is ±0.5 °C from 0 °C to 100 °C, and ±1.0 °C from −100 °C to 0 °C. If Std B had tolerances that are equal to these, or better, then it can be used as a substitute for Std A. By better, we refer to a more accurate tolerance, such as ±0.4 °C, ±0.25 °C, or ±0.1 °C. All of these tolerances would be considered to meet or exceed the tolerances of Std A.

Some calibration technicians substitute equipment based on manufacturer (like units should have the same tolerances, right? Wrong!), compatibility of ranges only (both standards have the same knobs and dials), or the ability of the substituted standard to replace the required standard because it was calibrated by the original standard to begin with (see circular calibration). Whatever the reason, caution must be taken when substituting standards, whether they are reference (used to calibrate your working standards) or working standards used to calibrate your customer's test equipment.

Loss of traceability will always come back to haunt any calibration function. We have all heard of Murphy's Law.[5] It will rear its ugly head at the most inopportune times, such as during an audit or inspection. Always maintain your lines of traceability and there will be no concern for substituting of reference or working standards.

The following are taken from NIST Special Publication 811, Sections 6 and 7. These Rules and Style Conventions can greatly assist anyone in his or her use and understanding of SI units.

Section 6: Rules and Style Conventions for Printing and Using Units

6.1 Rules and style conventions for unit symbols
The following eight sections give rules and style conventions related to the symbols for units.

6.1.1 Typeface
Unit symbols are printed in roman (upright) type regardless of the type used in the surrounding text. (See also Sec. 10.2 and Secs. 10.2.1 to 10.2.4.)

6.1.2 Capitalization
Unit symbols are printed in lower-case letters except that:
(a) the symbol or the first letter of the symbol is an upper-case letter when the name of the unit is derived from the name of a person; and
(b) the recommended symbol for the liter in the United States is L [see Table 6, footnote (*b*)].

Examples: m (meter) s (second) V (volt) Pa (pascal) lm (lumen) Wb (weber)

6.1.3 Plurals
Unit symbols are unaltered in the plural.

Example: l = 75 cm *but not*: l = 75 cms

Note: l is the quantity symbol for length. (The rules and style conventions for expressing the values of quantities are discussed in detail in Chapter 7.)

6.1.4 Punctuation
Unit symbols are not followed by a period unless at the end of a sentence.

Example: "Its length is 75 cm." or "It is 75 cm long." *but not*: "It is 75 cm. long. "

6.1.5 Unit symbols obtained by multiplication
Symbols for units formed from other units by multiplication are indicated by means of either a half-high (that is, centered) dot or a space. However, this *Guide*, as does Ref. [8], prefers the half-high dot because it is less likely to lead to confusion.

Example: N · m or Nm

Notes:
1. A half-high dot or space is usually imperative. For example, m · s^{-1} is the symbol for the meter per second while ms^{-1} is the symbol for the reciprocal millisecond (10^3 s^{-1}—see Sec. 6.2.3).
2. Reference [6: ISO 31-0] suggests that if a space is used to indicate units formed by multiplication, the space may be omitted if it does not cause confusion. This possibility is reflected in the common practice of using the symbol kWh rather than kW · h or kWh for the kilowatt hour. Nevertheless, this *Guide* takes the position that a half-high dot or a space should always be used to avoid possible confusion; and that for this same reason, only one of these two allowed forms should be used in any given manuscript.

6.1.6 Unit symbols obtained by division
Symbols for units formed from other units by division are indicated by means of a solidus (oblique stroke, /), a horizontal line, or negative exponents.

Example: m/s, $\dfrac{m}{s}$, or m · s^{-1}

However, to avoid ambiguity, the solidus must not be repeated on the same line unless parentheses are used.

Examples :   m/s^2 or m · s^{-2} *but not*: m/s/s
               m · kg/(s^3 · A) or m · kg · s^{-3} · A^{-1} *but not*: m · kg/s^3/A

Negative exponents should be used in complicated cases.

6.1.7 Unacceptability of unit symbols and unit names together
Unit symbols and unit names are not used together. (See also Secs. 9.5 and 9.8.)

Example: C/kg, C · kg^{-1}, or coulomb per *but not:* coulomb/kg; coulomb per kg; kilogram C/kilogram; coulomb · kg^{-1}; C per kg; coulomb/kilogram

6.1.8 Unacceptability of abbreviations for units
Because acceptable units generally have internationally recognized symbols and names, it is not permissible to use abbreviations for their unit symbols or names, such as sec (for either s or second), sq. mm (for either mm^2 or square millimeter), cc (for either cm^3 or cubic centimeter), mins (for either min or minutes), hrs (for either h or hours), lit (for either L or liter), amps (for either A or amperes), AMU (for either u or unified atomic mass unit), or mps (for either m/s or meter per second). Although the values of quantities are normally expressed using symbols for numbers and symbols for units (see Sec. 7.6), if for some reason the name of a unit is more appropriate than the unit symbol (see Sec. 7.6, note 3), the name of the unit should be spelled out in full.

6.2 Rules and style conventions for SI prefixes
The following eight sections give rules and style conventions related to the SI prefixes.

6.2.1 Typeface and spacing
Prefix symbols are printed in roman (upright) type regardless of the type used in the surrounding text, and are attached to unit symbols without a space between the prefix symbol and the unit symbol. This last rule also applies to prefixes attached to unit names.

Examples: mL (milliliter) pm (picometer) GΩ (gigaohm) THz (terahertz)

6.2.2 Capitalization
The prefix symbols Y (yotta), Z (zetta), E (exa), P (peta), T (tera), G (giga), and M (mega) are printed in uppercase letters while all other prefix symbols are printed in lowercase letters (see Table 5). Prefixes are normally printed in lowercase letters.

6.2.3 Inseparability of prefix and unit
The grouping formed by a prefix symbol attached to a unit symbol constitutes a new inseparable symbol (forming a multiple or submultiple of the unit concerned) which can be raised to a positive or negative power and which can be combined with other unit symbols to form compound unit symbols.

Examples: $2.3 \text{ cm}^3 = 2.3 \text{ (cm)}^3 = 2.3 \text{ (10}^{-2} \text{ m)}^3 = 2.3 \times 10^{-6} \text{ m}^3$
$1 \text{ cm}^{-1} = 1 \text{ (cm}^{-1}) = 1 \text{ (10}^{-2} \text{ m)}^{-1} = 10^2 \text{ m}^{-1}$
$5000 \text{ } \mu\text{s}^{-1} = 5000 \text{ (ms)}21 = 5000 \text{ (1026 s)}21 = 5 \text{ 3 } 109 \text{ s}21$
$1 \text{ V/cm} = (1 \text{ V)}/(10^{-2}\text{m}) = 10^2 \text{ V/m}$

Prefixes are also inseparable from the unit names to which they are attached. Thus, for example, millimeter, micropascal, and meganewton are single words.

6.2.4 Unacceptability of compound prefixes
Compound prefix symbols, that is, prefix symbols formed by the juxtaposition of two or more prefix symbols, are not permitted. This rule also applies to compound prefixes.

Example: nm (nanometer) *but not:* mμm (millimicrometer)

6.2.5 Use of multiple prefixes
In a derived unit formed by division, the use of a prefix symbol (or a prefix) in both the numerator *and* the denominator may cause confusion. Thus, for example, 10 kV/mm is acceptable, but 10 MV/m is often considered preferable because it contains only one prefix symbol and it is in the numerator.

In a derived unit formed by multiplication, the use of more than one prefix symbol (or more than one prefix) may also cause confusion. Thus, for example, 10 MV · ms is acceptable, but 10 kV · s is often considered preferable.

Note: Such considerations usually do not apply if the derived unit involves the kilogram. For example, 0.13 mmol/g is *not* considered preferable to 0.13 mol/kg.

6.2.6 Unacceptability of stand-alone prefixes
Prefix symbols cannot stand alone and thus cannot be attached to the number 1, the symbol for the unit one. In a similar vein, prefixes cannot be attached to the name of the unit one, that is, to the word *one*. (See Sec. 7.10 for a discussion of the unit one.)

Example: the number density of Pb atoms is $5 \times 10^6/m^3$ *but not:* the number density of Pb atoms is 5 M/m^3

6.2.7 Prefixes and the kilogram
For historical reasons, the name *kilogram* for the SI base unit of mass contains the name *kilo*, the SI prefix for 10^3. Thus, because compound prefixes are unacceptable (see Sec. 6.2.4), symbols for decimal multiples and submultiples of the unit of mass are formed by attaching SI prefix symbols to g, the unit symbol for gram, and the names of such multiples and submultiples are formed by attaching SI prefixes to the name *gram*.

Example: 10^{-6} kg = 1 mg (1 milligram) *but not:* 10^{-6} kg = 1 μkg (1 microkilogram)

6.2.8 Prefixes with the degree Celsius and units accepted for use with the SI
Prefix symbols may be used with the unit symbol °C and prefixes may be used with the unit name *degree Celsius*. For example, 12 m °C (12 millidegrees Celsius) is acceptable. However, to avoid confusion, prefix symbols (and prefixes) are not used with the time-related unit symbols (names) min (minute), h (hour), d (day); nor with the angle-related symbols (names) ° (degree), ' (minute), and " (second) (see Table 6).

Prefix symbols (and prefixes) may be used with the unit symbols (names) L (liter), t (metric ton), eV (electronvolt), and u (unified atomic mass unit) (see Tables 6 and 7). However, although submultiples of the liter such as mL (milliliter) and dL (deciliter) are in common use, multiples of the liter such as kL (kiloliter) and ML (megaliter) are not. Similarly, although multiples of the metric ton such as kt (kilometric ton) are commonly used, submultiples such as mt (millimetric ton), which is equal to the kilogram (kg), are not. Examples of the use of prefix symbols with eV and u are 80 MeV (80 megaelectronvolts) and 15 nu (15 nanounified atomic mass units).

Section 7: Rules and Style Conventions for Expressing Values of Quantities

7.1 Value and numerical value of a quantity
The *value* of a quantity is its magnitude expressed as the product of a number and a unit, and the number multiplying the unit is the *numerical value* of the quantity expressed in that unit.

More formally, the value of quantity A can be written as $A = \{A\} [A]$, where $\{A\}$ is the numerical value of A when the value of A is expressed in the unit $[A]$. The numerical value can therefore be written as $\{A\} = A / [A]$, which is a convenient form for use in figures and tables. Thus, to eliminate the possibility of misunderstanding, an axis of a graph or the heading of a column of a table can be labeled "*t*/°C" instead of "*t* (°C)" or "Temperature (°C)." Similarly, an axis or column heading can be labeled "*E*/(V/m)" instead of "*E* (V/m)" or "Electric field strength (V/m)."

Examples:
1. In the SI, the value of the velocity of light in vacuum is c = 299,792,458 m/s exactly. The number 299,792,458 is the numerical value of c when c is expressed in the unit m/s, and equals c /(m/s).
2. The ordinate of a graph is labeled $T/(10^3 \text{ K})$, where T is thermodynamic temperature and K is the unit symbol for kelvin, and has scale marks at 0, 1, 2, 3, 4, and 5. If the ordinate value of a point on

a curve in the graph is estimated to be 3.2, the corresponding temperature is $T/(10^3 \text{ K}) = 3.2$ or $T = 3200$ K. Notice the lack of ambiguity in this form of labeling compared with "Temperature (10^3 K)."

3. An expression such as $\ln(p /\text{MPa})$, where p is the quantity symbol for pressure and Mpa is the unit symbol for megapascal, is perfectly acceptable because p /MPa is the numerical value of p when p is expressed in the unit MPa and is simply a number.

Notes:
1. For the conventions concerning the grouping of digits, see Sec. 10.5.3.
2. An alternative way of writing $c /(\text{m/s})$ is $\{c\}_{m/s}$, meaning the numerical value of c when c is expressed in the unit m/s.

7.2 Space between numerical value and unit symbol
In the expression for the value of a quantity, the unit symbol is placed after the numerical value and a *space* is left between the numerical value and the unit symbol. The only exceptions to this rule are for the unit symbols for degree, minute, and second for plane angle: °, ', and ", respectively (see Table 6), in which case no space is left between the numerical value and the unit symbol.

Example: $\alpha = 30°22'8''$

Note: α a quantity symbol for plane angle.

This rule means that:
(a) The symbol °C for the degree Celsius is preceded by a space when one expresses the values of Celsius temperatures.

Example: $t = 30.2 \text{ °C}$ *but not:* $t = 30.2° \text{ C}$ or $t = 30.2°\text{C}$

(b) Even when the value of a quantity is used in an adjectival sense, a space is left between the numerical value and the unit symbol. (This rule recognizes that unit symbols are not like ordinary words or abbreviations but are mathematical entities, and that the value of a quantity should be expressed in a way that is as independent of language as possible—see Secs. 7.6 and 7.10.3.)

Examples: a 1 m end gauge *but not:* a 1-m end gauge
           a 10 kΩ resistor *but not:* a 10-kΩ resistor

However, if there is any ambiguity, the words should be rearranged accordingly. For example, the statement "the samples were placed in 22 mL vials" should be replaced with the statement "the samples were placed in vials of volume 22 mL."
Note: When unit names are spelled out, the normal rules of English apply. Thus, for example, "a roll of 35-millimeter film" is acceptable (see Sec. 7.6, note 3).

7.3 Number of units per value of a quantity
The value of a quantity is expressed using no more than one unit.

Example: $l = 10.234 \text{ m}$ *but not:* $l = 10 \text{ m } 23 \text{ cm } 4 \text{ mm}$

Note: Expressing the values of time intervals and of plane angles are exceptions to this rule. However, it is preferable to divide the degree decimally. Thus one should write 22.20° rather than 22°12', except in fields such as cartography and astronomy.

7.4 Unacceptability of attaching information to units
When one gives the value of a quantity, it is incorrect to attach letters or other symbols to the unit in order to provide information about the quantity or its conditions of measurement. Instead, the letters or other symbols should be attached to the quantity.

Example: $V_{max} = 1000 \text{ V}$ *but not:* $V = 1000 \text{ V}_{max}$

Note: V is a quantity symbol for potential difference.

7.5 Unacceptability of mixing information with units

When one gives the value of a quantity, any information concerning the quantity or its conditions of measurement must be presented in such a way as not to be associated with the unit. This means that quantities must be defined so that they can be expressed solely in acceptable units (including the unit one—see Sec. 7.10).

Examples:
the Pb content is 5 ng/L *but not:* 5 ng Pb/L or 5 ng of lead/L
the sensitivity for NO_3 molecules is 5×10^{10}/cm^3 *but not:* the sensitivity is 5×10^{10} NO_3 molecules/cm^3
the neutron emission rate is 5×10^{10}/s *but not:* the emission rate is 5×10^{10} n/s
the number density of O_2 atoms is 3×10^{18}/cm^3 *but not:* the density is 3×10^{18} O_2 atoms/cm^3
the resistance per square is 100 Ω *but not:* the resistance is 100 Ω/square

7.6 Symbols for numbers and units versus spelled-out names of numbers and units

This *Guide* takes the position that the key elements of a scientific or technical paper, particularly the results of measurements and the values of quantities that influence the measurements, should be presented in a way that is as independent of language as possible. This will allow the paper to be understood by as broad an audience as possible, including readers with limited knowledge of English. Thus, to promote the comprehension of quantitative information in general and its broad understandability in particular, values of quantities should be expressed in acceptable units using

- the Arabic symbols for numbers, that is, the Arabic numerals, *not* the spelled-out names of the Arabic numerals; and
- the symbols for the units, *not* the spelled-out names of the units.

Examples:
the length of the laser is 5 m *but not:* the length of the laser is five meters
the sample was annealed at a temperature of 955 K for 12 h *but not:* the sample was annealed at a temperature of 955 kelvins for 12 hours

Notes:
1. If the intended audience for a publication is unlikely to be familiar with a particular unit symbol, it should be defined when first used.
2. Because the use of the spelled-out name of an Arabic numeral with a unit symbol can cause confusion, such combinations must strictly be avoided. For example, one should never write "the length of the laser is five m."
3. Occasionally, a value is used in a descriptive or literary manner and it is fitting to use the spelled-out name of the unit rather than its symbol. Thus this *Guide* considers acceptable statements such as "the reading lamp was designed to take two 60-watt light bulbs," or "the rocket journeyed uneventfully across 380,000 kilometers of space," or "they bought a roll of 35-millimeter film for their camera."
4. The *United States Government Printing Office Style Manual* (Ref. [4], pp. 165–171) gives the rule that symbols for numbers are always to be used when one expresses (a) the value of a quantity in terms of a unit of measurement, (b) time (including dates), and (c) an amount of money. This publication should be consulted for the rules governing the choice between the use of symbols for numbers and the spelled-out names of numbers when numbers are dealt with in general.

7.7 Clarity in writing values of quantities

The value of a quantity is expressed as the product of a number and a unit (see Sec. 7.1). Thus, to avoid possible confusion, this *Guide* takes the position that values of quantities must be written so that it is completely clear to which unit symbols the numerical values of the quantities belong. Also to avoid possible confusion, this *Guide* strongly recommends that the word "to" be used to indicate a range of values for a quantity instead of a range dash (that is, a long hyphen) because the dash could be misinterpreted as a minus sign. (The first of these recommendations once again recognizes that unit symbols are not like ordinary words or abbreviations but are mathematical entities—see Sec. 7.2.)

Examples:
51 mm x 51 mm x 25 mm *but not:* 51 $\times$ 51 $\times$ 25 mm
225 nm to 2400 nm or (225 to 2400) nm *but not:* 225 to 2400 nm
0 °C to 100 °C or (0 to 100)°C *but not:* 0 °C − 100 °C
0 V to 5 V or (0 to 5) V *but not:* 0 − 5 V
(8.2, 9.0, 9.5, 9.8, 10.0) GHz *but not:* 8.2, 9.0, 9.5, 9.8, 10.0 GHz
63.2 m ± 0.1 m or (63.2 ± 0.1) m *but not:* 63.2 ± 0.1 m or 63.2 m ± 0.1
129 s − 3 s = 126 s or (129 − 3) s = 126 s *but not:* 129 − 3 s = 126 s

Note: For the conventions concerning the use of the multiplication sign, see Sec. 10.5.4.

7.8 Unacceptability of stand-alone unit symbols
Symbols for units are never used without numerical values or quantity symbols (they are not abbreviations).

Examples:
there are 10^6 mm in 1 km *but not:* there are many mm in a km
it is sold by the cubic meter *but not:* it is sold by the m^3
t/°C, E /(V/m), p /MPa, and the like are perfectly acceptable (see Sec. 7.1)

7.9 Choosing SI prefixes
The selection of the appropriate decimal multiple or submultiple of a unit for expressing the value of a quantity, and thus the choice of SI prefix, is governed by several factors. These include:

the need to indicate which digits of a numerical value are significant
the need to have numerical values that are easily understood
the practice in a particular field of science or technology

A digit is significant if it is required to express the numerical value of a quantity. In the expression $l = 1200$ m, it is not possible to tell whether the last two zeroes are significant or only indicate the magnitude of the numerical value of l. However, in the expression $l = 1.200$ km, which uses the SI prefix symbol for 10^3 (kilo, symbol k), the two zeroes are assumed to be significant because if they were not, the value of l would have been written $l = 1.2$ km.

It is often recommended that, for ease of understanding, prefix symbols should be chosen in such a way that numerical values are between 0.1 and 1000, and that only prefix symbols that represent the number 10 raised to a power that is a multiple of 3 should be used.

Examples:
3.3×10^7 Hz may be written as 33×10^6 Hz = 33 MHz
0.009 52 g may be written as 9.52×10^{-3} g = 9.52 mg
2703 W may be written as 2.703×10^3 W = 2.703 kW
5.8×10^{-8} m may be written as 58×10^{-9} m = 58 nm

However, the values of quantities do not always allow this recommendation to be followed, nor is it mandatory to try to do so.

In a table of values of the same kind of quantities or in a discussion of such values, it is usually recommended that only one prefix symbol should be used even if some of the numerical values are not between 0.1 and 1000. For example, it is often considered preferable to write "the size of the sample is 10 mm $\times$ 3 mm $\times$ 0.02 mm" rather than "the size of the sample is 1 cm $\times$ 3 mm $\times$ 20 μm." In certain kinds of engineering drawings it is customary to express all dimensions in millimeters. This is an example of selecting a prefix based on the practice in a particular field of science or technology.

7.10 Values of quantities expressed simply as numbers: the unit one, symbol 1

Certain quantities, such as refractive index, relative permeability, and mass fraction, are defined as the ratio of two mutually comparable quantities and thus are of dimension one (see Sec. 7.14). The coherent SI unit for such a quantity is the ratio of two identical SI units and may be expressed by the numeral 1. However, the numeral 1 generally does not appear in the expression for the value of a quantity of dimension one. For example, the value of the refractive index of a given medium is expressed as $n = 1.51 \times 1 = 1.51$.

On the other hand, certain quantities of dimension one have units with special names and symbols which can be used or not depending on the circumstances. Plane angle and solid angle, for which the SI units are the radian (rad) and steradian (sr), respectively, are examples of such quantities (see Sec. 4.3).

7.10.1 Decimal multiples and submultiples of the unit one

Because SI prefix symbols cannot be attached to the unit one (see Sec. 6.2.6), powers of 10 are used to express decimal multiples and submultiples of the unit one.

Example: $\mu_r = 1.2 \times 1026$ *but not:* $\mu_r = 1.2\ \mu$

Note: μ_r is the quantity symbol for relative permeability.

7.10.2 %, percentage by, fraction

In keeping with Ref. [6: ISO 31-0], this *Guide* takes the position that it is acceptable to use the internationally recognized symbol % (percent) for the number 0.01 with the SI and thus to express the values of quantities of dimension one (see Sec. 7.14) with its aid. When it is used, a space is left between the symbol % and the number by which it is multiplied [6: ISO 31-0]. Further, in keeping with Sec. 7.6, the symbol % should be used, not the name "percent."

Example: $x_B = 0.0025 = 0.25\%$ *but not:* $x_B = 0.0025 = 0.25\%$ or $x_B = 0.25$ percent

Note: x_B is the quantity symbol for amount-of-substance fraction of B (see Sec. 8.6.2).

Because the symbol % represents simply a number, it is not meaningful to attach information to it (see Sec. 7.4). One must therefore avoid using phrases such as "percentage by weight," "percentage by mass," "percentage by volume," or "percentage by amount of substance."

Similarly, one must avoid writing, for example, "% (*m/m*)," "% (by weight)," "% (*V/V*)," "% (by volume)," or "% (mol/mol)." The preferred forms are "the mass fraction is 0.10," or "the mass fraction is 10%," or "$w_B = 0.10$," or "$w_B = 10\%$" (w_B is the quantity symbol for mass fraction of B—see Sec. 8.6.10); "the volume fraction is 0.35," or "the volume fraction is 35%," or "$\varphi_B = 0.35$," or "$\varphi_B = 35\%$" (φ_B is the quantity symbol for volume fraction of B—see Sec. 8.6.6); and "the amount-of-substance fraction is 0.15," or "the amount-of-substance fraction is 15%," or "$x_B = 0.15$," or "$x_B = 15\%$." Mass fraction, volume fraction, and amount-of-substance fraction of B may also be expressed as in the following examples: $w_B = 3$ g/kg; $\varphi_B = 6.7$ mL/L; $x_B = 185\ \mu$mol/mol. Such forms are highly recommended. (See also Sec. 7.10.3.)

In the same vein, because the symbol % represents simply the number 0.01, it is incorrect to write, for example, "where the resistances R_1 and R_2 differ by 0.05%," or "where the resistance R_1 exceeds the resistance R_2 by 0.05%." Instead, one should write, for example, "where $R_1 = R_2(1 + 0.05\%)$," or define a quantity Δ via the relation $\Delta = (R_1 - R_2)/R_2$ and write "where $\Delta = 0.05\%$." Alternatively, in certain cases, the word *fractional* or *relative* can be used. For example, it would be acceptable to write "the fractional increase in the resistance of the 10 kΩ reference standard in 1994 was 0.002%."

7.10.3 ppm, ppb, and ppt

In keeping with Ref. [6: ISO 31-0], this *Guide* takes the position that the language-dependent terms part per million, part per billion, and part per trillion, and their respective abbreviations "ppm," "ppb," and "ppt" (and similar terms and abbreviations), are not acceptable for use with the SI to express the values of quantities. Forms such as those given in the following examples should be used instead.

Examples:
a stability of 0.5 (μA/A)/min *but not:* a stability of 0.5 ppm/min
a shift of 1.1 nm/m *but not:* a shift of 1.1 ppb
a frequency change of 0.35 3 10^{-9} *f but not:* a frequency change of 0.35 ppb
a sensitivity of 2 ng/kg *but not:* a sensitivity of 2 ppt

the relative expanded uncertainty of the resistance *R* is $U_r = 3$ μΩ/Ω; or the expanded uncertainty of the resistance *R* is $U = 3 \times 10^{-6}$ *R;* or the relative expanded uncertainty of the resistance *R* is *U* r = 3×10^{-6} *but not:* the relative expanded uncertainty of the resistance *R* is $U_r = 3$ ppm

Because the names of numbers 10^9 and larger are not uniform worldwide, it is best that they be avoided entirely (in most countries, 1 billion = 1×10^{12}, not 1×10^9 as in the United States); the preferred way of expressing large numbers is to use powers of 10. This ambiguity in the names of numbers is one of the reasons why the use of ppm, ppb, ppt, and the like is deprecated. Another, and a more important one, is that it is inappropriate to use abbreviations that are language dependent together with internationally recognized signs and symbols, such as MPa, ln, 10^{13}, and %, to express the values of quantities and in equations or other mathematical expressions (see also Sec. 7.6).

Note: This *Guide* recognizes that in certain cases the use of ppm, ppb, and the like may be required by a law or a regulation. Under these circumstances, Secs. 2.1 and 2.1.1 apply.

7.10.4 Roman numerals
It is unacceptable to use Roman numerals to express the values of quantities. In particular, one should not use C, M, and MM as substitutes for 10^2, 10^3, and 10^6, respectively.

7.11 Quantity equations and numerical-value equations
A quantity equation expresses a relation among quantities. An example is *l = vt*, where *l* is the distance a particle in uniform motion with velocity *v* travels in the time *t*.

Because a quantity equation such as *l = vt* is independent of the units used to express the values of the quantities that compose the equation, and because *l*, *v*, and *t* represent quantities and not numerical values of quantities, it is incorrect to associate the equation with a statement such as "where *l* is in meters, *v* is in meters per second, and *t* is in seconds."

On the other hand, a numerical value equation expresses a relation among numerical values of quantities and therefore does depend on the units used to express the values of the quantities. For example, $(l)_m = 3.6^{-1} (v)_{km/h} (t)_s$ expresses the relation among the numerical values of *l*, *v*, and *t* only when the values of *l*, *v*, and *t* are expressed in the units meter, kilometer per hour, and second, respectively. (Here $(A)_x$ is the numerical value of quantity *A* when its value is expressed in the unit X—see Sec. 7.1, note 2.)

An alternative way of writing the above numerical value equation, and one that is preferred because of its simplicity and generality, is $l/m = 3.6^{-1}$ [*v* /(km/h)](*t*/s). NIST authors should consider using this preferred form instead of the more traditional form "*l* = 3.6^{-1} *vt*, where *l* is in meters, *v* is in kilometers per hour, and *t* is in seconds." In fact, this form is still ambiguous because no clear distinction is made between a quantity and its numerical value. The correct statement is, for example, "*l** = 3.6^{-1} *v* t**, where *l** is the numerical value of the distance *l* traveled by a particle in uniform motion when *l* is expressed in meters, *v** is the numerical value of the velocity *v* of the particle when *v* is expressed in kilometers per hour, and *t** is the numerical value of the time of travel *t* of the particle when *t* is expressed in seconds." Clearly, as is done here, it is important to use different symbols for quantities and their numerical values to avoid confusion.

It is the strong recommendation of this handbook that, because of their universality, quantity equations should be used in preference to numerical-value equations. Further, if a numerical value equation is used, it should be written in the preferred form given in the above paragraph and, if at all feasible, the quantity equation from which it was obtained should be given.

Notes:
1. Two other examples of numerical-value equations written in the preferred form are as follows, where E_g is the gap energy of a compound semiconductor and κ is the conductivity of an electrolytic solution:

$E_g/\text{eV} = 1.425 - 1.337x + 0.270x^2$, $0 \leq x \leq 0.15$, where x is an appropriately defined amount-of-substance fraction (see Sec. 8.6.2)
$\kappa/(\text{S/cm}) = 0.065\ 135 + 1.7140 \times 10^{-3}(t/^\circ C) + 6.4141 \times 10^{-6}(t/^\circ C)^2 - 4.5028 \times 10^{-8}(t/^\circ C)^3$, $0\ ^\circ C \leq t \leq 50^\circ C$, where t is Celsius temperature

2. Writing numerical-value equations for quantities expressed in inch-pound units in the preferred form will simplify their conversion to numerical-value equations for the quantities expressed in units of the SI.

7.12 Proper names of quotient quantities
Derived quantities formed from other quantities by division are written using the words "divided by" rather than the words "per unit" in order to avoid the appearance of associating a particular unit with the derived quantity.

Example: pressure is force divided by area *but not:* pressure is force per unit area

7.13 Distinction between an object and its attribute
To avoid confusion, when discussing quantities or reporting their values, one should distinguish between a phenomenon, body, or substance, and an attribute ascribed to it. For example, one should recognize the difference between a body and its mass, a surface and its area, a capacitor and its capacitance, and a coil and its inductance. This means that although it is acceptable to say "an object of mass 1 kg was attached to a string to form a pendulum," it is not acceptable to say "a mass of 1 kg was attached to a string to form a pendulum."

7.14 Dimension of a quantity
Any SI derived quantity Q can be expressed in terms of the SI base quantities length (l), mass (m), time (t), electric current (l), thermodynamic temperature (T), amount of substance (n), and luminous intensity (l_v) by an equation of the form

$$Q = l^\alpha m^\beta t^\gamma l^\delta T^\varepsilon n^\zeta l_v^\eta \sum_{k=1}^{K} a_k,$$

where the exponents α, β, γ, . . . are numbers and the factors a_k are also numbers. The dimension of Q is defined to be

$$\dim Q = L^\alpha\ M^\beta\ T^\gamma\ I^\delta\ \Theta^\varepsilon N^\zeta\ J^\eta\ ,$$

where L, M, T, I, Θ, N, and J are the *dimensions* of the SI base quantities length, mass, time, electric current, thermodynamic temperature, amount of substance, and luminous intensity, respectively. The exponents α, β, γ, . . . are called *dimensional exponents*. The SI derived unit of Q is $m^\alpha \cdot kg^\beta \cdot s^\gamma \cdot A^\delta \cdot K^\varepsilon \cdot mol^\zeta \cdot cd^\eta$, which is obtained by replacing the dimensions of the SI base quantities in the dimension of Q with the symbols for the corresponding base units.

Example: Consider a nonrelativistic particle of mass m in uniform motion which travels a distance l in a time t. Its velocity is $v = l/t$ and its kinetic energy is $E_k = m\ v^2/2 = l^2 m t^{-2}/2$. The dimension of E_k is $\dim E_k = L^2\ MT^{-2}$ and the dimensional exponents are 2, 1, and -2. The SI derived unit of E_k is then $m^2 \cdot kg \cdot s^{-2}$, which is given the special name *joule* and special symbol J.

A derived quantity of dimension one, which is sometimes called a *dimensionless quantity*, is one for which all of the dimensional exponents are zero: dim $Q = 1$. It therefore follows that the derived unit for such a quantity is also the number one, symbol 1, which is sometimes called a *dimensionless derived unit*.

Example: The mass fraction w_B of a substance B in a mixture is given by $w_B = m_B/m$, where m_B is the mass of B and m is the mass of the mixture (see Sec. 8.6.10). The dimension of w_B is dim $w_B = M^1 M^{-1} = 1$; all of the dimensional exponents of w_B are zero, and its derived unit is $kg^1 \cdot kg^{-1} = 1$ also.

NOTES

1. "BIPM—Bureau International des Poids et Measures. Metrology and World Metrology Day," June 2006. http://www.bipm.org/en/practical_info/faq/welcome.html (18 August 2006).

2. Jay L. Bucher, *The Metrology Handbook* (Milwaukee: ASQ Quality Press, 2004), chapter 11.

3. B. N. Taylor, "Guide for the Use of the International System of Units (SI)," (March 2004). http://physics.nist.gov/Pubs/SP811/contents.html (18 August 2006).

4. Pelican Products, http://www.pelican.com/ (21 August 2006).

5. "Murphy's Law," 11 August 2006. http://en.wikipedia.org/wiki/Murphy's_law (21 August 2006). Murphy's law (also known as Finagle's law or Sod's law) is a popular adage in Western culture that broadly states that things will go wrong in any given situation in which error is possible. "If there's more than one way to do a job, and one of those ways will result in disaster, then somebody will do it that way." It is most commonly formulated as "Anything that can go wrong will go wrong." Technically speaking, this latter definition is incorrect, given that it refers more accurately to the law of pessimism, Finagle's Law. In American culture the law was named after Major Edward A. Murphy, Jr., a development engineer working for a brief time on rocket sled experiments done by the United States Air Force in 1949.

8

Traceability

There are three different yet connected subjects that should be discussed in this chapter. The first, of course, is traceability, or the unbroken chain of comparisons back to a national or international standard. The second is reverse traceability and the third is circular calibration.

Before discussing traceability, the definition must first be established. According to the *International Vocabulary of Basic and General Terms in Metrology* (VIM), ISO, 1993, the definition of *traceability* is "property of the result of a measurement or the value of a standard whereby it can be related to stated references, usually national or international standards, through an unbroken chain of comparisons all having stated uncertainties."

For calibration practitioners, the important phase to remember is "an unbroken chain of comparisons—or calibrations." In the metrology community, which is where calibration practitioners ply their trade, an unbroken chain of comparisons is usually called the traceability chain, or paper trail.

The reason *stated uncertainties* is not referred to is simple. It is assumed that all statements, which are usually written in the form of a calibration certificate, calibration record, or in the calibration procedure, will have the stated uncertainties stated or recorded somewhere within them. It makes no difference if the statement is a formula, an equation, or a simple policy of TUR $\geq$ 4:1 (Test uncertainty ratio is equal to or greater than four to one). The uncertainty is known and this is part of the unbroken chain of comparisons.

A certain percentage of calibration practitioners have never heard of the paper trail because they have never been involved in audits of any kind or their function does not fall under any type of quality system. It is important to understand how a paper trail is set up and how to show a complete paper trail when being audited. Let's start with the standard used for performing calibration and work up the chain to a national standard, and then back down the chain to the item being calibrated.

In theory, the standard that is used to calibrate your test equipment should be either a reference standard (if you are calibrating working standards) or a working standard. If it is a reference standard, it should be sent out to a third-party calibration lab on a regular basis. When the reference standard is received back after calibration, it must have a calibration certificate.

It behooves the user who receives the calibration certificate to check not only to ensure the standard, or item sent out for calibration, is the same as the one on the certificate, but also that the item works properly after being received from the vendor. It is not uncommon to receive the correct standard back from the vendor, but have the wrong certificate. This is the first thing to check.

The second is if the standard was calibrated against a higher standard that had a TUR of equal to or greater than 4:1. The third item to check is if the standard passed its initial calibration, or *As Found* readings. Information on not passing the *As Found* calibration can be found in Chapter 6.

Let's assume that the correct certification is returned with the standard. How do you know if the uncertainty of the standard used to calibrate your item is sufficient for your needs?

According to 17025, paragraphs 5.10.2 and 5.10.4:

> Each certificate of calibration shall include:
> 1. A title
> 2. Name and address of the laboratory
> 3. Unique identification of the certificate
> 4. Name and address of the client
> 5. Identification of the method used
> 6. A description and condition of the item calibrated
> 7. The date(s) of calibration
> 8. The calibration results and units of measurement
> 9. The name, function, and signature of the person authorizing the certificate
> 10. The environmental conditions during calibration
> 11. The uncertainty of measurement
> 12. Evidence that the measurements are traceable.[1]

Usually the third-party vendor will have a statement along the lines of "Uncertainties have been estimated at a 95% confidence level (k=2)" or "Calibration at a TUR of 4:1 provides reasonable confidence that the unit under test (UUT) is within the stated tolerances on the certificate or manufacturer's published specifications." These meet the conditions of number eleven (11) above. Some vendors go into great detail in listing the uncertainties, while others use statements that meet the minimum requirements.

How does all this fit into your quality calibration program? Because internal calibration departments usually do not generate calibration certificates (not a requirement and too costly in time and materials), they have to still meet the requirements of the paper trail. This is accomplished by using the calibration record and calibration label attached to the item that is calibrated. Here is where the discussion of the paper trail from the calibrated instrument comes into play.

At a minimum, a third-party calibration laboratory used by any company to calibrate their standards (or any other pieces of test equipment) should be certified to ISO 17025. The quality of calibrations performed, and their paper trail are only as good as each link in that chain. And one of the links is the third-party calibration lab used by a company. The reference or working standard is used to calibrate the company's various pieces of test equipment. A calibration record is generated, and in that record is proof that traceability has occurred. This is done by having a statement in the calibration record explaining the traceability chain. There must be something referring the calibration traceability back to a national or international standard. The standards used should be listed in the record, along with when they are next due calibration (it had better be after the date the calibration was performed). During a review, the auditor only has to follow when the test instrument was calibrated, to when the standard was calibrated, to the calibration certificate of the standard showing traceability back to the national or international standard. An unbroken chain of comparisons, and your auditor is satisfied, and your calibrations are repeatable and traceable . . . with the documentation to prove it.

Reverse traceability usually comes into play when a standard, rather a reference or working standard, is found to be out of tolerance either during the normal calibration cycle, when making a routine check, or if there is a question about the results of a measurement. If a standard is found to be out of tolerance, there are certain functions that must be performed.

If a reference or working standard was found to be out of tolerance during normal calibration, and adjusted to again meet specifications, then it can still be used as a standard. If it is found to be out of tolerance, or not working correctly during the normal course of the day, it needs to be immediately removed from service and segregated from other standards or labeled as unusable. It is critical to identify the unit as unusable to prevent further problems.

In either case, all items that were calibrated using that standard must be identified and recalibrated using a known good standard. This is where reverse traceability comes into play. A quality calibration system must have a way to show what items were calibrated, during a specific period of time (example: since the last time the standard was calibrated) so they can be recalled and/or identified. A risk assessment must be made to determine if the items calibrated by the out-of-tolerance standard need to be recalled or if there is no impact on them. This is one of the reasons that both *As Found* and *As Left* readings are required when performing calibrations within your department or sending standards out for third-party calibrations. Analysis of the *As Found* readings will help determine the extent of *damage* an out-of-tolerance standard has caused.

The damage could fall at either end of a spectrum, or usually somewhere in between. If the standard was out of tolerance a minor amount and the uncertainty calculations showed that by using it previously all calibrations still maintained their TUR of 4:1 (the standard could be so accurate that all previous calibration had a TUR of 10:1), then it is possible that nothing needs to be done other than making sure the historical record reflects what happened.

On the other end of the spectrum, every item calibrated using that standard has to be recalled and a risk assessment of all production using the recalled test equipment is also suspect. Hopefully, it is something in between.

The important question is this: Does your software have the ability to provide reverse traceability on the standards that you use? Can items calibrated using those standards be readily identified in case of recall. Even the most basic quality system should have the ability to do this, even if it means examining every calibration record to see what standard was used to perform all the calibrations to see if a bad standard was used. This can be very time consuming and labor intensive. Having reverse traceability as part of your software program just in case the inevitable happens can save time, money, and a lot of headaches on down the road.

It is also important to have the ability to select a date range for reverse traceability. One must be able to focus on the calibration that occurred after the last calibration, because it is assumed that calibrations made before that were good, because the standard came back within tolerance during that previous calibration.

For example, standard XYZ was sent out for calibration on February 1, 2005. When it returned, the *As Found* readings were within tolerance, so no *As Left* readings were required. The standard was checked to ensure it was not damaged in shipment and that it was making good readings against another standard to check for accuracy and reliability. Standard XYZ was put back into service as a working or reference standard with a calibration date of February 4, 2005, and a date due calibration of February 4, 2006 (a one-year calibration interval).

Standard XYZ was sent out again for calibration on February 2, 2006 (shortly before coming due for calibration on the 4th). It was found to be out of tolerance by the third-party calibration laboratory, and notification was given to the user. At this point, the user should be performing a risk analysis because they have now been notified of a problem, and should also know what the *As Found* readings were. (In some industries, but not all, it is critical to be notified immediately of out-of-tolerance standards. With other companies, notification when the standard and calibration certificates are returned is soon enough. It is up to the user of the standard to make this determination.) Reverse traceability needs to be performed for all calibrations using standard XYZ during the period of February 1, 2005, to the present. Prior to February 1, 2005, it is assumed it was working correctly and within tolerance because it was received by the outside vendor within tolerance on that date. However, even though the standard was checked and verified upon its return from calibration, it is not known when it went out of tolerance, so that is why the recall dates start at the last calibration.

Some industries recognize this problem and try to be proactive by using check standards of some type to assure themselves, and their customers, that their standards are at least working close to their prescribed tolerances. The check standards can be used each

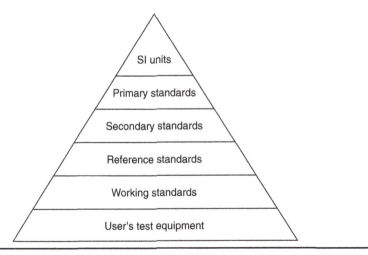

Figure 8.1 Traceability pyramid.

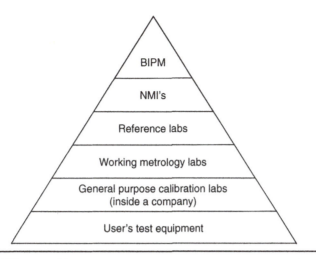

Figure 8.2 Traceability pyramid.

time the standard is turned on, or during a prescribed time frame such as daily, weekly, or monthly. The check standard could be a simple go–no go gage or device. Or there could be a requirement to do a partial calibration against a more accurate standard. No matter the device, being proactive is usually more beneficial than being reactive when it comes to reverse traceability or product recalls. A standard traceability pyramid is similar to the one in Figures 8.1 and 8.2.

NIST POLICY ON TRACEABILITY[2]

NIST has developed an organizational policy on traceability and a set of related supplementary materials, which includes answers to questions frequently asked by customers of NIST measurement services. The policy and supplementary materials are intended to serve as a resource for NIST customers.

Introduction

The mission of NIST is to develop and promote measurement, standards, and technology to enhance productivity, facilitate trade, and improve the quality of life. To help meet the measurement and standards needs of U.S. industry and the nation, NIST provides calibrations, standard reference materials, standard reference data, test methods, proficiency evaluation materials, measurement quality assurance programs, and laboratory accreditation services that assist a customer in establishing traceability of *results of measurements* or *values of standards*.

Traceability requires the establishment of an *unbroken chain of comparisons* to *stated references*. NIST assures the traceability of results of measurements or values of standards that NIST itself provides, either directly or through an *official NIST program or collaboration*. Other organizations are responsible for establishing the traceability of their own results or values to those of NIST or other stated references. NIST has adopted this policy statement to document the NIST role with respect to traceability.

Statement of Policy

To support the conduct of its mission and to ensure that the use of its name, products, and services is consistent with its authority and responsibility, NIST:

1. Adopts for its own use and recommends for use by others the definition of traceability provided in the most recent version of the *International Vocabulary of Basic and General Terms in Metrology:* "property of the result of a measurement or the value of a standard whereby it can be related to stated references, usually *national* or *international standards,* through an unbroken chain of comparisons all having *stated uncertainties.*" (International Vocabulary of Basic and General Terms in Metrology [VIM], BIPM, IEC, IFCC, ISO, IUPAC, IUPAP, OIML, 2nd ed., 1993, definition 6.10)
2. Establishes traceability of the results of its own measurements and values of its own standards and of results and values provided to customers in NIST calibration and measurement certificates, operating in accordance with the *NIST System for Assuring Quality in the Results of Measurements Delivered to Customers in Calibration and Measurement Certificates.* See http://www.nist.gov/nistsystem/.
3. Asserts that providing support for a claim of traceability of the result of a measurement or value of a standard is the responsibility of the *provider* of that result or value, whether that provider is NIST or another organization; and that assessing the validity of such a claim is the responsibility of the *user* of that result or value.
4. Communicates, especially where claims expressing or implying the contrary are made, that NIST does not *define, specify, assure,* or *certify* traceability of the results of measurements or values of standards except those that NIST itself provides, either directly or through an official NIST program or collaboration. (See also NIST Administrative Manual, Subchapter 5.03, *NIST Policy on Use of its Name in Advertising* at http://ts.nist.gov/traceability/503.htm)
5. Collaborates on development of standard definitions, interpretations, and recommended practices with *organizations that have authority and responsibility for variously defining, specifying, assuring, or certifying traceability.*
6. Develops and disseminates technical information on traceability and conducts coordinated outreach programs on issues of traceability and related requirements.
7. Assigns responsibility for oversight of implementation of the NIST policy on traceability to the NIST Measurement Services Advisory Group.

The third subject for this chapter is circular calibration. This is an occurrence when there is no chain of comparisons to a high-level standard. Consider this example. A master set of gage blocks is used to calibrate a company's batch of micrometers that are used as working standards. The calibration technicians use any one of these micrometers to calibrate the company's various parts and products. Once a year, the master set of gage blocks is checked using one of the micrometers that had the highest resolution (in this case, erroneously thought to mean it is more accurate than the other micrometers). However, the micrometer that is used to check the master set of gage blocks is itself checked against the gage blocks once a year for its normal calibration. As can be seen, there is no unbroken chain of comparisons to a higher level standard. In this case, the chain of comparisons goes in a circle, hence the term circular calibration. The master set of gage blocks needs to be sent out on a regular basis to be calibrated against a more accurate standard, which in turn has been calibrated to a higher level or national/ international standard—an unbroken chain of comparisons.

Will there always be a need for traceability? There can be no doubt. With the world growing smaller and smaller, literally and figuratively speaking, the need has become more critical than ever. How is the world growing smaller? As the world population continues to crowd our land masses, and with global warming slowly melting the ice caps; as the oceans slowly rise reducing the already overcrowded land masses, we are slowly running out of space. So, our world is literally growing small.

With the globalization of industry, it has become even more critical that measurements and accuracies be the same everywhere trade, commerce, and manufacturing occur. The traceability of our measurements to a central standard is even more important as we go into the future. In the year 2006, when someone reads about traceability to a national or international standard, they probably assume that in the United States that it refers automatically to NIST. There may come a day when traceability refers to one central International Metrology Institute (IMI instead of NMI) to ensure we are all really singing from the same page. However, my crystal ball was broken many years ago, so this is only a glance into a future that may or may not come true.

NOTES

1. ANSI/ISO/IEC, *ANSI/ISO/IEC 17025-2005: General requirements for the competence of testing and calibration laboratories* (Milwaukee: ASQ Quality Press, 2005).

2. "NIST Policy on Traceability," 3 December 2001. http://ts.nist.gov/traceability/nist%20traceability %20policy-external.htm (21 August 2006).

9

Uncertainty

W hen a manufacturer produces test equipment, they specify certain parameters that must be met (for example, an operating range of $-10\ °C$ to $50\ °C$). Why then does the unit need to be calibrated in an environmentally controlled laboratory when it is not used there? In most cases it doesn't! An environment that controls temperature, humidity, vibration, dust, and so on reduces the uncertainty of measurement, and allows for the calibration of state-of-the-art test equipment in a calibration laboratory, instead of at NIST, by a third-party calibration laboratory, or by the manufacturer. If a company follows the guidance in ANSI/NCSL Z540-1-1994, paragraph 10.2(b): *"The laboratory shall ensure that the calibration uncertainties are sufficiently small so that the adequacy of the measurement is not affected. Well-defined and documented measurement assurance techniques or uncertainty analyses may be used to verify the adequacy of a measurement process. If such techniques or analyses are not used, then the collective uncertainty of the measurement standards shall not exceed 25% of the acceptable tolerance (that is, manufacturer's specification) for each characteristic of the measuring and test equipment being calibrated or verified"*; they will have traceable measurement.

This 25% of the acceptable tolerance is where the 4:1 ratio is derived. Here's an example: your standard has an accuracy of 0.25 °C; the most accurate item you can calibrate could be no better than 1.0 °C, or four times less accurate. You must ensure that the combined uncertainty of the standard(s) being used is at least four times more accurate than what is being calibrated. By calibrating test equipment in the environment where it is normally used, you can determine how well it works and the repeatability of that instrument. This is also where traceability is derived. Traceability is defined as *an unbroken chain of comparisons* back to an international/national standard. During each calibration along that chain, a 4:1 ratio, or uncertainty analysis must be documented. Only then can you show your equipment is traceable back to a specific standard or level of accuracy.

However, one must keep in mind that this is the minimum TUR. Anything smaller, 3:1, 2:1, and so on, would probably require actually doing uncertainty analysis using an uncertainty calculator or possibly one of the formulas that follow.

In order to combine the uncertainty components, one method that can be used is the *root sum square (RSS) method.*

$$u_c = \sqrt{X_1^2 + X_2^2 + X_3^2 + \ldots + X_n^2}$$

Table 9.1 Confidence interval values and associated k values.

Coverage Factor (k)	Confidence Level
1.000	68.27
1.645	90.00
1.960	95.00
2.000	95.45
2.576	99.00
3.000	99.73

Combined uncertainty is assigned a coverage factor k that denotes a degree of confidence interval associated with it. The GUM (guide to the expression of uncertainty in measurement) recommends k=2 at 95% confidence interval. *Combined uncertainty* multiplied by the *coverage factor: k* is known as *expanded uncertainty* and denoted *U*.

$$U = k \cdot u_c$$

Various confidence interval values and their associated k values are shown in Table 9.1.

MEASUREMENT SYSTEM ANALYSIS—MSA[1]

Measurement system analysis (MSA) is an experimental and mathematical method of determining how much the variation within the measurement process contributes to overall process variability.

There are five parameters to investigate in an MSA: bias, linearity, stability, repeatability, and reproducibility. According to Automotive Industry Action Group (AIAG), a general rule of thumb for measurement system acceptability is:

- Under 10% error is acceptable.
- Ten percent to 30% error suggests that the system is acceptable depending on the importance of application, cost of measurement device, cost of repair, and other factors.
- Over 30% error is considered unacceptable; you should improve the measurement system.

AIAG also states that the number of distinct categories the measurement systems divides a process into should be greater than or equal to five. In addition to percent error and the number of distinct categories, you should also review graphical analyses over time to decide on the acceptability of a measurement system.[1] MSA is one of the core components of the Six Sigma approach. Many problems encountered with Statistical Process Control (SPC) and Design of Experiments (DOE) are caused by problems with measurement systems.

Measurement Systems Analysis (Gage R&R Studies) is also a core component of the QS-9000/TS 16949 Automotive Quality System Requirements. The standard is supported by manuals published by the AIAG. The MSA standard is described in the *Third Edition of the Measurement Systems Analysis Manual,* published in March 2002. I will briefly touch on one of the areas in MSA because it is generally industry-specific and not usually addressed in other industry calibration environments.

MSA examines the nature of measurement systems and the fundamental analyses used to examine them. Sources of variation can be found by analyzing a measuring device's bias, linearity, stability, repeatability, and reproducibility. However, it can be said that measurement system errors can be classified into two general categories: precision and accuracy.

By definition, *precision* is a property of a measuring system or instrument. Precision is a measure of the repeatability of a measuring system—how much agreement there is within a group of repeated measurements of the same quantity under the same conditions. Precision is *not* the same as *accuracy*.

Also by definition, *accuracy of a measurement* is a qualitative indication of how closely the result of a measurement agrees with the true value of the parameter being measured. Because the true value is always unknown, *accuracy* of a measurement is always an estimate. An accuracy statement by itself has no meaning other than as an indicator of quality. It has quantitative value only when accompanied by information about the uncertainty of the measuring system.

Accuracy of a measuring instrument (different than accuracy of a measurement) is a qualitative indication of the ability of a measuring instrument to give responses close to the true value of the parameter being measured. Accuracy is a design specification and may be verified during calibration.

When referred to in the automotive industry, it could be said that accuracy describes the difference between the measurement and the part's actual value. Precision describes the variation you see when you measure the same part repeatedly with the same device or instrument. Within a measurement system you can have one or both of these problems . . . accuracy and/or precision. To illustrate, see the circles in Figure 9.1. You can have test equipment that measures parts precisely but not accurately (second circle from the left). You could also have test equipment that is accurate but not precise (third circle from the left). At the far end of the scale, you could possibly have test equipment that is neither accurate nor precise (last circle on the right). Of course the object is to be precise as well as accurate, and that is illustrated by the first circle on the left.

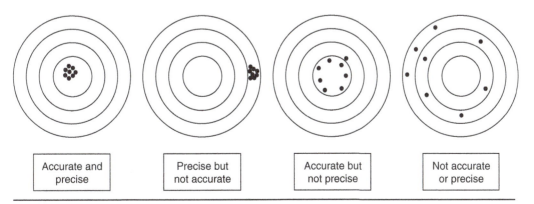

| Accurate and precise | Precise but not accurate | Accurate but not precise | Not accurate or precise |

Figure 9.1 Accuracy versus precision.

Earlier in the book it was explained how the calibration of one item would not be shown, but the calibration of all items would. I would like to deviate from that plan on one specific type of equipment: balances.

If the use of balances is not required in the daily work environment, this portion of this chapter may be skipped. However, there are alternative ways to find uncertainty explained that may come in handy in other areas of the work environment.

In 1997, when I was trying to find the specifications for the balances used where I was employed, it was found that none were listed anywhere, other than the *U.S. Pharmacopeia*. To easily understand what was written there, at the time, was beyond my understanding. Simply put, if the directives were to be fully complied with they would 1) each require extensive testing, calculations, and documentation to find their accuracies; and 2) only about half of each balance's range could be used for any type of traceable weighing.

On the first count, there was not enough time while putting a quality system in place to do all the testing and calculating; and on the second count there was no way the company was only going to use half the range of all the balances that had been previously purchased, and purchase more to cover the lower ranges that would not be covered. There had to be an alternative approach that satisfied both the requirements of ISO 9000, ISO 13485, and cGMP, while at the same time meeting the budget constraints of the company.

I hope you can follow the chain of calculations and summations that follow. These may not work for every company, but they will hopefully give the reader an idea of how necessity (the mother of invention) stirred me to look for alternative methods to solve traceability and budget restraint problems.

According to the *U.S. Pharmacopeia*; Apparatus for Tests and Assays; <41> Weights and Balances (p. 1869): *Unless otherwise specified, when substances are to be "accurately weighed" for Assay the weighing is to be performed with a weighing device whose measurement uncertainty (random plus systematic error) does not exceed 0.1% of the reading.*

The following shows examples of what 0.1% of certain masses would equate to in the real world:

Examples:	Reading:	0.1% Rdg
	1 Kg	1 gram
	100 g	100 mg
	10 grams	10 mg
	1 gram	1 mg
	.1 gram	.1 mg

Table 9.2 shows the class of weight required to meet the above tolerances. As can be seen, the smaller the value, the higher accuracy class required to meet the specifications.

The following shows what the *U.S. Pharmacopeia* says should be done to find the lowest denomination that would be able to be weighed on a balance or scale. According to the *U.S. Pharmacopeia*; Apparatus for Tests and Assays; <41> Weights and Balances (p. 1869): . . . (finding the lowest mass that can be weighed:) *Measurement uncertainty is satisfactory if three times the standard deviation of not less than ten replicate weighings divided by the amount weighed, does not exceed 0.001. . . .*

Table 9.2 Weight requirements to meet prescribed tolerances.

Nominal Value	ASTM Class 1	ASTM Class 2	ASTM Class 3	ASTM Class 4
1 mg	0.01 mg	0.014 mg	0.025 mg	0.05 mg
2 mg	0.01 mg	0.014 mg	0.025 mg	0.05 mg
3 mg	0.01 mg	0.014 mg	0.026 mg	0.052 mg
5 mg	0.01 mg	0.014 mg	0.028 mg	0.055 mg
10 mg	**0.01 mg**	0.014 mg	0.6 mg	0.1 mg
20 mg	0.01 mg	**0.014 mg**	0.035 mg	0.7 mg
50 mg	0.01 mg	0.014 mg	**0.042 mg**	0.085 mg
100 mg	0.01 mg	0.025 mg	0.5 mg	**0.1 mg**
200 mg	0.01 mg	0.025 mg	0.6 mg	0.12 mg
300 mg	0.01 mg	0.025 mg	0.7 mg	0.14 mg
500 mg	0.01 mg	0.025 mg	0.8 mg	0.16 mg
1 g	0.034 mg	0.054 mg	0.1 mg	0.2 mg
10 g	0.05 mg	0.074 mg	0.25 mg	0.5 mg

The balance used for this experiment in Table 9.3 reads to five decimal places. Its calculations according to the *U.S. Pharmacopeia* are listed. In this example, the weight under test is 10 milligrams. To have satisfactory measurement uncertainty, then the calculated value of three times the standard deviation divided by the weight (0.0000172879 milligrams $\times$ 3/10 milligrams = 0.000005186 milligrams) cannot exceed 0.001 $\times$ 10 milligrams, or 0.00001 milligrams. And in this case it does not.

Under the *U.S. Pharmacopeia,* generally speaking, the following are guidelines for masses to be used as minimums for weighing:

4 decimal places:	200 mg ~ 300 mg
5 decimal places:	50 mg
6 decimal places:	7.5 mg

This is just a general rule. As can be seen in the example of Table 9.3, 10 milligrams could be weighed on that particular balance and still meet the requirements of the *U.S.*

Table 9.3 Example for finding the lowest mass that can be weighed.

9.99999	Computation for 5 decimal places
9.99997	Using 10 mg, class-1 weight
9.99994	
9.99997	
9.99997	
9.99994	
9.99994	
9.99997	
9.99995	
9.99997	
	0.000005186
9.999961	weight
1.72884E-05	standard deviation
5.18654E-06	3 times standard deviation divided by weight

Pharmacopeia. Should every company that falls under the guidelines of the *U.S. Pharmacopeia* do these calculations for each and every one of their balances? It is hard to tell. They must determine the accuracy and tolerances they can live with in the use of their balances. However, reducing the uncertainty of its balances (not uncalibrating . . . two different things) a company may find that it can still use the balances on hand, while traceable measurements are made throughout its processes.

As stated earlier, balance manufacturers do not give specific accuracy or uncertainty statements about their balances. They also do not list the bottom end of where a balance can accurately be used. The customer must figure this out. The reasoning behind this is simple, manufacturers do not know what kind or type of weight standards each customer may or may not use. They do not want to be incriminated for inaccurate measurements or results of those measurements if something goes wrong at the customer's end. Generally, accuracy of any type found in balance manuals or on the manufacturer's websites are for non-linearity only.

The following is an example of how to calculate balance uncertainties without having to lose some of their usable range and accuracy.

An Optional Way to Figure Range and Tolerances for Balances

Balance XYZ: 100 g

Readability:	0.01 mg
Repeatability:	0.01 mg
Non-Linearity:	0.03 mg

(A) Find the least accurate of Repeatability, Readability, and Linearity: 0.03 mg
(B) Double (0.06 mg), and add one major count: 0.07 mg
(C) Increase this number (0.07 mg) to the next significant number of 1 or 5 (10, 50, 100, 500, and so on) if not already a 1 or 5: Uncertainty = ± 0.1 mg

The process for calculating the lowest mass that can be weighed:

(D) Multiply (C) *(0.1 mg)* by 10 *(1 mg)*. *To err on the side of caution and compensate for weighing uncertainties (operator, drafts, rounding, temperature, eccentric loading, hysteresis, humidity, and vibration), we increase by a factor of 10.*

The process for calculating usable range:

(E) Multiply the balance's uncertainty (C) by 1000 (the reverse, 0.1% of reading). This is the starting point for 0.1% of reading *(0.1 mg* × 1000 = 0 .1 grams)
Multiply (D) and go to the number calculated in (E) for low range ± the accuracy *(1 mg ~ 0.1 grams,* ± 0.1 mg)
The high range is from (E) to the balance's top weighable limit *(0.1 grams ~ 100* grams, ± 0.1% Rdg)

The following is how a limited calibration label might read on the balance from the previous example:

Unless otherwise stated on a limited calibration label, this balance has an accuracy of

① ± .1 mg (1 mg to .1 g) and
② > .1 g ± 0.1% of Reading.

The usable range of this balance is: 1 mg ~ 100 g

Comparing different methods of calculating uncertainty:

Using the specifications from the same balance, keeping in mind that there is uncertainty in the repeatability, but not in readability (the use of the least significant digit [LSD] would come into play with the readability).

Readability:	0.01 mg
Repeatability:	0.01 mg
Linearity:	± 0.03 mg

RSS (root sum squared) method:

$$U = \sqrt[2]{(0.01)^2 + (0.03)^2}$$
$$U = \sqrt[2]{(0.0001) + (0.0009)}$$
$$U = \sqrt[2]{.001}$$
$$U = 2 \cdot 0.0216227$$
$$U = 0.0632454$$
$$U = 0.063 \, \text{mg} (k = 2)$$

RSS with Swiss modification method:

$$U = \sqrt[2]{(0.01)^2 + \tfrac{2}{3}(0.03)^2}$$
$$U = \sqrt[2]{(0.0001) + (0.0006)}$$
$$U = \sqrt[2]{.0007}$$
$$U = 2 \cdot 0.0264575$$
$$U = 0.052915$$
$$U = 0.053 \, \text{mg} (k = 2)$$

RSS with Hart Scientific modification*:

$$U = \sqrt[2]{\left(\frac{0.01}{\sqrt{3}}\right)^2 + \left(\frac{0.03}{\sqrt{3}}\right)^2}$$
$$U = \sqrt[2]{(0.0000333) + (0.0003)}$$
$$U = \sqrt[2]{.0003333}$$
$$U = 2 \cdot 0.0182565$$
$$U = 0.036513$$
$$U = 0.037 \, \text{mg} (k = 2)$$

* An alternative approach takes advantage of offsetting errors by converting *peak-to-peak* specifications to approximate standard deviations (by dividing them by $\sqrt{3}$), adding the components using the RSS method, and multiplying by 2 to give 95% coverage.

4:1 ratio (adding uncertainties):

U = .01 + .03
U = 2 · .04
U = 0.08 *mg* (k = 2)

Alternative method:

U = 0.03
U = 2 · 0.03
U = 0.06 + 1 *digit*
U = 0.07
U = 0.1 *mg*

The result 0.03 comes from finding the least accurate of Readability, Repeatability, and Linearity. The final value comes from doubling the value of 0.03, adding one count (or digit—LSD), then increasing that number to the next significant number of 1 or 5 (unless the number is already a 1 or a 5).

Computation comparisons:

RSS:	0.063 mg
RSS (Swiss mod):	0.053 mg
RSS (Hart-Scientific):	0.037 mg
4 to 1:	0.08 mg
Alternative method:	0.1 mg

As can be seen in the comparison chart, the uncertainty increases (gets worse), while the computation method becomes easier. Please keep in mind we are dealing with balances used in a biotechnology or pharmaceutical environment. It has been found that unless extremely accurate readings are a requirement, that generally to err on the side of caution—increased uncertainty, while getting the most usable and accurate range out of your balances, is just plain good business practice.

The following is calculated using Tolerance Calculator:

Evaluating the effect of different TURs on consumer and producers' risk for a 95% confidence level

1V Nominal with ± 2% Tolerance

Nominal	Tol.	Tol. Minus	Tol. Plus	C Risk	P Risk	TUR	Req Uncert	Conf.
1V	0.02	0.98	1.02	0.41	0.53	10:1	0.002	95%
1V	0.02	0.98	1.02	0.87	1.6	4:1	0.005	95%
1V	0.02	0.98	1.02	1.06	2.36	3:1	0.006667	95%
1V	0.02	0.98	1.02	1.35	4.31	2:1	0.01	95%
1V	0.02	0.98	1.02	1.81	13.39	1:1	0.02	95%

UNCERTAINTY CALCULATOR[2]

Uncertainty Calculator (UC) was written to provide a straightforward means for computing and documenting measurement uncertainty analysis. Since its introduction as a freeware software application in 1997 UC has gone through several revisions (the latest revision is 3.2). Since its release, user registrations have been submitted from over 54 countries spanning major governmental and commercial test and calibration facilities as well as privately owned mom and pop shops. UC works with Windows platforms from Windows 95 to XP and can be installed on a server for distributed computing. UC is available for download from various websites such as:

Agilent Metrology Forum	http://metrologyforum.tm.agilent.com/download3.shtml
Cal Lab Magazine	http://www.callabmag.com/freeware.html
T&MW Magazine	http://www.reed-electronics.com/tmworld/index.asp
	?layout=siteinfo&doc_id=61299
eCalibration.com	http://www.ecalibration.com/News/2002/April/
	products8.htm

UC is provided as a self-extracting executable (.exe) file. You can download the .exe file to your desktop and click on it to install. The default directory installation for UC is C:\Program Files\Uncertainty Calculator 3.2. This directory will contain the application UnCal3.2.exe as well as the following;

UC_Readme.rtf	Installation instructions (Rich text file)
UC_Manual.rtf	UC user manual (Rich text file)
UC_Tutorial.xls	User tutorial (Excel file)

In addition there are 26 example files; each starting with UC-. These examples span from RF Power to Mass and are obtained from various publications, which are cited in each example. Clicking on UnCal3.2.exe will invoke the UC opening screen shown in Figure 9.2.

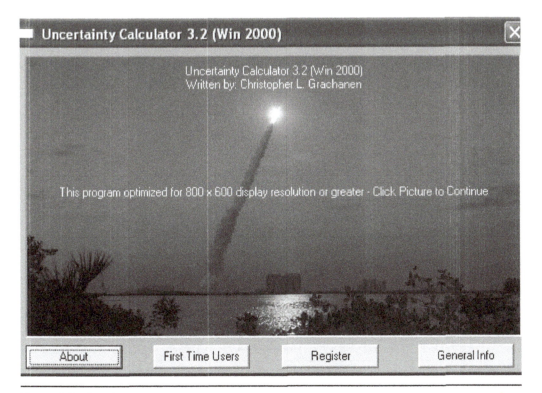

Figure 9.2   Opening screen.

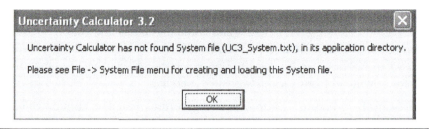

Figure 9.3   First-time user screen 1.

From the opening screen, users can print the registration fax form and obtain general setup and reference information about UC. Clicking anywhere on the picture will start the UC main program. The first time UC is invoked, two pop-ups will be displayed (this is normal). See Figures 9.3 and 9.4.

UC employs two user-created files: A System file to identify paths to Windows applications, a necessity for different Windows versions, and an Options file to save user preferences.

SYSTEM FILE

The System file is accessed from the UC main application drop down menu File/System file. From the System file screen, clicking the *Read Me* button located at the top right of the screen will provide the following information. (Note: the following step numbers correspond to the red numbers appearing on the System file screen)

Steps for Entering Application Paths into System File

1. Select "Info Type" to APPLICATIONS.
2. Select an Application, such as Write, Word, and so on, via "Application Name."
3. Locate path for application selected in Step 2 via "Current Drive / Directories."
4. Select an application path from Step 3 via "Copy Current Path."
5. Export the selected path from Step 4 to System File via "Export Copied Path."
6. Repeat steps 2–5 for remainder of applications.
7. Save the System File via "Save System File."

For a typical Windows XP desktop with Office 2003 installed, the following paths are applicable: (Note: Application paths may be determined via Windows Search.)

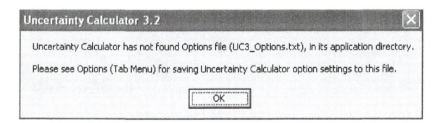

Figure 9.4   First-time user screen 2.

Write	C:\WINDOWS\system32\
Word	C:\Program Files\Microsoft Office\OFFICE11\
Excel	C:\Program Files\Microsoft Office\OFFICE11\
Data Files	May be any local directory or network directory
Windows Calculator	C:\WINDOWS\system32\
User Defined #1	May be any local directory or network directory
User Defined #2	May be any local directory or network directory
UC Manual	C:\Program Files\Uncertainty Calculator 3.2\
UC Tutorial	C:\Program Files\Uncertainty Calculator 3.2\

Note: User Defined 1 and 2 are applications that can be invoked directly from UC, that is, Tolerance Calculator, Mismatch Uncertainty Calculator, and so on, by providing the path (the name of the application's .exe file is entered in *Entering User Defined Information*).

Hint: It is recommended that users set up subdirectories for each uncertainty parameter, that is, AC Voltage, Torque, Microwave, and so on, and set up only the main directory path for the Data Files path entry. This makes for easier file retrieval and provides a logical format for creating UC's summary files. UC's summary file control, located in File/Data Files menu, creates summary files from individual UC Uncertainty files, listing key information from each file (a convenient means to summarize laboratory uncertainties without having to copy and paste info from each individual file).

Steps for Entering User-Defined Information

User-defined information fields are used to enter text about the facility performing the uncertainty analysis that will appear on UC-generated reports as well as user defined application names.

1. Select "Info Type" to USER.
2. Enter company info for reports (use preceding spaces to center info on reports).
3. Enter user application filenames. Example: TolCal3.exe.
4. Click "Update" to copy entered info to System File.
5. Follow steps 1, 4–7 above for entering user application paths (Step 2.b).

Options File

UC Options file may be accessed via the Options tab menu from the top of the UC main screen. Default selections are initially checked. To save selections click the Save button located at the upper right portion of the Options tab menu.

Uncertainty Analysis Data Entry

The following is the automatic screen sequence for data entry (Auto Sequence ON) from UC's main screen:

1. Select an Active Line Type (Type A or B) and line number (or click on Type A or B grid line).
2. Enter a description about the uncertainty component; <ENTER>.
3. Enter an uncertainty value or an asterisk "*" and Std Uncertainty; <ENTER>.
4. Enter or select an uncertainty unit; <ENTER>.

5. Select entries for Distribution Screen; <OK>.
6. Repeat steps 1 through 5 for each Type A and B uncertainty component.
7. Compute Effective Degrees of Freedom, if applicable (See Screens Menu).
8. Compute Coverage Factor (See Screens Menu).
9. <COMPUTE EXPANDED UNCERTAINTY> to calculate Expanded Uncertainty.

Note: To enter data manually, set "Auto Sequence Entries" to OFF (Unchecked) and enter data directly into a data entry field (click a data entry header box to invoke a data entry screen).

Additional Information

UC has an extensive Users manual (rich text file) as well as a tutorial (Excel), accessible from the Files_Info_Tools tab menu. These documents reside in UC's application directory and can be opened using Windows Explorer. UC also has an extensive Help menu accessible by clicking on any question mark (?) appearing on a white dialog icon.

WHAT DOES THE CRYSTAL BALL SHOW FOR THE FUTURE?

Where might uncertainty be in 5, 10, or 20 years? There can be no doubt that we will always need to know what the tolerances are for any given piece of test equipment and standard used to calibrate it. It seems that the biggest detractor to uncertainty budgets entering the 21st century is the apprehension of performing complicated math, or taking the time to analyze the data and plug the numbers into a computer program.

There may come a day when an item's uncertainty is programmed into its bar code label, or retrieved from a universal data base for inclusion into any company's CAMS. A universal service bus (USB) thumb drive or permanent web link with the range and tolerances in a special formula for use in uncertainty budgets could accompany every purchase of test equipment, similar to today's items coming with CDs.

Manufacturers, calibration laboratories, metrology departments, and test equipment users will have the final say in determining the viability and usage of uncertainty budgets. As the old adage goes: "The squeaky wheel gets the grease," so it will go with uncertainty budgets. Simplicity will be the driving factor, once the masses realize that the need for knowing the uncertainty will always be with us, but how to determine the end result does not have to be brain surgery or rocket science any longer.

NOTES

1. Automotive Industry Action Group (AIAG), "Measurement Systems Analysis Reference Manual. Chrysler, Ford, General Motors Supplier Quality Requirements Task Force." 29 August 2003. http://www.isixsigma.com/dictionary/Measurement_System_Analysis_-_MSA-277.htm (22 August 2006).

2. Uncertainty Calculator (rev. 3.2 released March 2002). A FREEWARE application developed and produced by Chris Grachanen of Hewlett-Packard for Windows 2000 (including Win98, NT, and so on). Uncertainty Calculator 3.2 addresses uncertainties for commonly made measurements in a simple, straightforward manner; congruent with the basic guidelines contained within measurement uncertainty publications such as ISO "Guide to the Expression of Uncertainty in Measurement (GUM)," NIST Technical Note 1297, and so on. Some features of the Uncertainty Calculator 3.2 software program are numerical methods for computing sensitivity coefficients, methodology for computing correlated input quantities, an equation writer that facilitates creating and documenting measurement equations, a system file, and the Welch-Satterthwaite equation for computing effective degrees of freedom.

10

Calibration Labels

What is the status of your test equipment? Does the user know? Can an auditor tell? Do you have to go to your computer or get a printout to know? The proper use of calibration labels and their reflection in your calibration management system could be the answer to all these questions. Quick, simple, and easy to use; they are a requirement in most systems.

ANSI/ISO/ASQ Q10012-2003 reflects the need for labels as stated in paragraph 7.1.1, General, which reads: "Information relevant to the metrological confirmation status of measuring equipment shall be readily available to the operator, including any limitations or special requirements." ANSI/ASQC M1-1996 states in paragraph 4.10, Identification of Calibration Status: "Instruments shall be labeled to indicate when the next calibration is due." And Calibration Control Systems for the Biomedical and Pharmaceutical Industry, RP-6 reads in paragraph 5.8, Labels: "To alert the user to the status of a piece of equipment, all equipment should be labeled or coded to indicate its status. Equipment not labeled is considered 'not calibrated.'"[1]

It should be written into a quality calibration system that any time a piece of test equipment does not have a calibration label, it is assumed that it is not calibrated and is not to be used. Any item that is calibrated would have a calibration label affixed to it that would be easy to see and identify as such. If the label has fallen off or been placed in an obscure or hidden area, or if the calibration technician simply forgot to affix a label after calibration, the item still cannot be used until the proper calibration sticker has been placed on the item, or the item has been recalibrated. The reasoning behind such a policy is simple; it is a bulletproof policy that removes any doubt from the customer and the calibration practitioners. No label, the item is not calibrated. If CAMS has been updated properly and a calibration record can be found that matches CAMS, then a new calibration label can be placed on the unit. No doubt, no problem. But all of the above-stated items must be in place first—CAMS data that matches a calibration record. Without both, a full calibration must be performed before the item can be used.

Sometimes equipment is of a particular size or shape that does not allow for the labels to be easily attached. In some of those cases, color codes are used to show when they are next due for calibration, by monthly schedules, quarterly schedules, and so on. In those cases, the system that is followed should be well documented and there is a procedure that identifies which color is for which time period, and so on.

When an actual date is used on a calibration label, everyone, including the user and calibration technician, knows that the piece of test equipment can be used till midnight on the date it is due for calibration. There is no vacillation in when it is due, or when it can be no longer used. It is in black and white and leaves no room for error or compromise. That is the way a quality calibration system should be developed and maintained.

A less-preferred method that is still used is a system that allows calibration to be performed during a specific month, week, or any other broad time frame. The reasons behind this are twofold. First, not having a specific date on the calibration label gives the user and calibration technician the impression that a piece of test equipment has some sort of expandability in its usability. This is not true. Calibration intervals (see Chapter 15) are designed to keep the test equipment in the field for as long as possible without going past their reliability date. By giving a time frame of ± 30 days, this reliability is compromised.

The following is an example of how using monthly dates can compromise your reliability. Let's assume the unit we are referring to has a calibration interval of 30 days, or 1 month. It was last calibrated on February 1. It is next due calibration during the month of March. According to this system, it can be calibrated any time between March 1 and March 31, without being considered overdue for calibration. If the unit is calibrated on March 1, it has met the requirement and has been in use for 28 days since the last calibration. If it is calibrated on March 31, it still meets the requirements but has been in use for 62 days. It is now due during April. If it is calibrated on April 5, it has been in use for five days (if calibrated on March 31). The company is not getting their money's worth in this type of system. Data collection of repeatability of the item is almost nonexistent since the time frames between calibration are so erratic. Also, the more times a calibration technician performs a calibration on a piece of test equipment, the costlier that piece of test equipment becomes. Needless to say, using a day, month, year time frame for your calibration intervals has all the benefits of a quality calibration system and none of the pitfalls of the alternatives.

Shown in Figure 10.1 are standard calibration labels, both large and small sizes. The DDC in the small label stands for Date Due Calibration. Both labels show the minimum required information by most standards. As can be seen, the only difference is the size. Generally, the larger label would be used on most test equipment. However, there are many types of test equipment that will not accommodate a large calibration label. That is when you would use the smaller label.

If the test equipment is such that there is no room for any size label, or due to how the item is used, such as a total immersion thermometer, there are alternative ways to show the calibration status of the item.

Calibration date _____

By _____

Next due _____

ID# _____

Cal. date _____
By _____
ID# _____
DDC _____

Figure 10.1   Samples of calibration labels.

Some of the options are placing a calibration label on a small manila tag and affixing the tag to the test instrument. Another option is the use of small metal tags, which are also attached to the test equipment.

In rare cases where there is no option for attaching a calibration label, due to size, how the item is used, or the environment where the item is used, showing the status of the item's calibration is still possible. Some organizations must use a binder to keep their calibration labels. Each calibration label is, of course, married to a particular piece of test equipment through the use of a unique identifier, which is on the test equipment, the label, and used in CAMS. In some cases, such as thermometers, the serial number is etched into the unit and that serial number must also be either written on the label or used in some way to easily match the calibration label to the test instrument. It is critical to keep the calibration label binder where it can be easily referred to when checking on the calibration status of any piece of test equipment.

In some cases an entire system is given one identification number, even though there might be more than one piece of test equipment in the system. This is allowed only if the entire system is calibrated as one unit. If each piece receives its own calibration, then each piece in the system must have its own identification number, calibration label, and be listed by itself in CAMS. If each item in a system that has a single identifier can be worked on separately, or receive adjustment or repair separate from the whole system, then each piece in the system should have its own identification number, calibration label, and be listed in CAMS on its own. It is important to be able to retrieve data, both from calibrations and repairs as needed on each item. And it can be very helpful to sort information by types of equipment. If some like items are listed within a system and not by themselves, the organization may loose valuable data when looking for trends about a particular item.

There are various reasons why a piece of test equipment might require a limit calibration. The standards required to give a full calibration are not available, so only part of the unit can be calibrated to its full specifications. The standards used to calibrate the unit might not have the required TUR or tolerances to meet or exceed 4:1, therefore the unit must be decertified to a lesser tolerance to maintain the TUR of greater than 4:1. It is also possible that not all of the ranges or parameters of an item are working properly or to the required specifications, and those ranges or parameters will not be calibrated. There are other reasons, depending on the industry or usage of the equipment. In any case, the test equipment must then receive a limited calibration label.

Two examples of limited calibration labels can be found in Figure 10.2. The border on the left and the diagonal line on the right would be red in color.

The label has the same information as a standard calibration label, with the addition of an area to list the limits imposed on the test instrument. In some cases, there might not be a limited calibration, but the user must be informed that a calibration chart or some other data is available to be used with the test equipment. This should be listed on a limit calibration label. The main idea of a limit calibration label is to draw the attention of the user to the label. That is why there might be a red border on the label or a red diagonal line across the label.

If there is not enough room on a standard limited calibration label to list all of the ranges or parameters that are not in use, then it is possible to list only the ranges or

| ID# _____ By _____ |
| Date Cal'd. _____ |
| Next DDC_____ |
| Limits: _____ |

| ID# _____ By _____ |
| Date Cal'd. _____ |
| Next DDC_____ |
| Limits: _____ |

Figure 10.2 Samples of limited calibration labels.

parameters that are working properly. Also, the limited space could be used to inform the user of the test equipment to find additional information in a binder, or other appropriate data used with the test instrument, and possibly the location of that data.

The important thing to remember when using limited calibration labels is that the user can understand the limits, tolerances, or available ranges or parameters of the test equipment involved. There is no use affixing a limited calibration label if the user cannot decode your intentions. The limits should be clear, concise, and readable.

Most quality systems specifically state that a piece of test equipment must have a calibration label affixed to it or the item is considered to be out of calibration. That does not always equate to the user knowing the status of a piece of test equipment. In most cases, as long as the item turns on, and performs in a similar manner as it did the previous day, the user is willing to bet that it is within calibration and will continue to use the item. This is where a label similar to the one in Figure 10.3 can be used (the background might be red to draw attention to it).

A *Do Not Use* label, similar to the two examples in Figure 10.3, can be used to draw attention to the fact that the piece of test equipment is out of calibration or out of service. In either case, it cannot be used by the customer under any circumstance.

In most cases, if the calibration technician removes the calibration label and affixes a *Do Not Use* label, several things have been accomplished. First, without the calibration label on the unit, nobody can say it was thought to be still calibrated. Some people leave the calibration label on the unit even though they have attached the *Do Not Use* label. Second, some technicians believe they need the data on the calibration label to complete the next calibration record. All the data on the calibration label should be available in CAMS, or the calibration technician can use the data from the removed calibration label

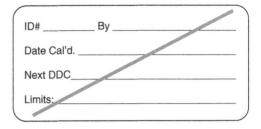

Figure 10.3 Samples of *Do Not Use* labels.

when they update records showing they have attached the *Do Not Use* label. It never hurts to have too much information in CAMS or your records.

As with most labels used by calibration technicians, any of the labels mentioned here can be produced in-house if needed and customized to meet the needs of the Metrology or Calibration Department. There is nothing in the regulations or standards that requires a commercially produced label be used. The only requirements are what must be on the label as a minimum. If a company is supporting a large database of test equipment, it might be more fiscally responsible to purchase labels in bulk. But for the seldom used or rare label, a printer and sticky-backed labels might do just as well. Such is the case with the label in Figure 10.4. For commercial labels, the ID number can still be written on it, even though there is no formal space. This does not invalidate the label; it only adds to its usefulness since it identifies the test equipment that the label should be affixed to.

Most organizations use some type of *Calibration Not Required* label if the test equipment does not require calibration. It precludes looking for a calibration label since it shows no calibration is needed. The label in Figure 10.4 (which might have red letters to draw attention) has an additional line for identifying the unit that it is affixed to. Why should that matter? On the rare occasion that some users are not as honest as one would hope, it can keep honest people honest. As an example, if an item requires calibration, but has gone overdue, it has been known during an audit for the user to simply pull off a *Calibration Not Required* sticker and affix it to the calibrated item (actually, over the top of the calibration label). Once the auditor has left, the *Calibration Not Required* sticker can be removed and nobody is the wiser. By have an area for placing an identification number, this might keep them honest since the test instrument's ID number should be prominently showing on the test instrument.

Two different yet similar labels are shown in Figure 10.5. The first, *Standardize Before Use,* is usually attached to items that require some sort of standardization before the item can be used. Examples of this are pH meters and conductivity meters. The user must standardize a probe with the meter using some type of buffer or solution. Usually the buffer or solution is traceable to NIST or some other standard. By using the buffer or solution, the probe and meter are *married* together and the user can determine if they are working properly by reading a slope or predetermined number. The label is used as a reminder to accomplish the standardization before using the instrument. Some organizations allow the unit to be standardized either once a day or within a few hours' time of each use. If that is the case, then there should be a logbook or binder kept that is filled in by the user

Figure 10.4 Samples of *Calibration Not Required* labels.

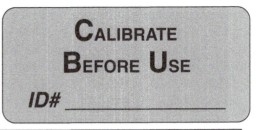

Figure 10.5   Sample *Standardize Before Use* and *Calibrate Before Use* labels.

to show when standardization was accomplished, and record what the slope or reading was at the time.

The second label, *Calibrate Before Use* (CBU), has been used in the military for many years. There are occasions when a piece of test equipment is rarely used, or used only during certain exercises, or with certain types of equipment that are also seldom used. A CBU sticker allows the organization to calibrate the item before it is used, and affix a calibration label to the unit. Once the item goes past its calibration due date, it can no longer be used, but is not considered overdue for calibration. It simply sits on the shelf till the next time it is needed, then the calibration technician is called upon to calibrate it again. After the calibration is performed another calibration label is affixed, and it goes back into a calibrated state until the calibration interval runs out. This saves the company time in calibrating an item that is seldom used, but still critical to the operation of the organization. These items are the exception to the rule, and an organization must be careful in selecting those items that they wish to code as CBU.

Calibration labels have not changed much over the years. They require a unique test equipment identifier, the date calibration was accomplished, the next date it is due for calibration, and a way to know who performed the calibration—whether it is by name, signature, or some uniquely identified stamp or number.

Wouldn't it be nice if in the future calibration labels were printed on special paper that allowed changes that were linked to time or date? This paper could be activated by heat, electricity, light, or any other form that provided a catalyst for starting the clock.

Once a predetermined time has passed, say six months or one year, the data on the label could disappear, or the label could turn red, or a large red X would appear across the label. This way the user would automatically be notified that they should not use that particular piece of test equipment. That is not to say that there are those that would not use it anyway, but by drawing attention to the label, it would at least ensure the user was notified that the item was out of calibration. Most users have other things on their mind when using their test equipment, and looking at the calibration label each and every time it is used is generally not one of those things they are thinking about.

There could come a time when the use of nanotechnology could be incorporated into the calibration label. Once the overdue date is reached, if the unit is plugged into a wall, the fuse could automatically blow, or the unit be incapacitated in some way that did not allow for its use until it was recalibrated and a new clock started.

Wishful thinking on the part of an experienced calibration technician? Not really. If the non-use of a piece of test equipment would keep disasters, poisonings, or accidents from happening, it won't be that long before the measures are put in place to assist managers and supervisors in preventing the use of out-of-calibration test equipment. It is far cheaper to be proactive in prevention than to be reactive in cleaning up the mess. Examples have been given in Chapter 1.

NOTE

1. Jay L. Bucher, *The Metrology Handbook* (Milwaukee: ASQ Quality Press, 2004), 95.

Part III

Developing a Quality Calibration Program

11

Meeting Your Needs
and Requirements

Each company, metrology department, or calibration function will have specific requirements that they have to meet, depending on which standard or regulation they
follow. No matter the size of the department or standard/regulation that must be
met, there are certain facets of a quality calibration program that must be in place and
working for them to benefit the organization.

When decisions are being made as to which pieces of test equipment should be identified and placed into a program, a couple of factors need to be considered. Does the test
equipment require calibration? Will the test equipment be adjusted, aligned, and repaired
in-house, or sent out for vendor support work? No matter which way things go, will there
be a need to record the work completed, whether in-house or by a vendor? If the answer
is yes to any of these questions, then the piece of test equipment must be identified, given
a unique ID number, and placed in the quality calibration system. This is accomplished
through labeling and database management. As previously mentioned, this computerized
system will be called CAMS (Computer Automated Management System; more can be
found on CAMS in Chapter 14).

Within any CAMS, a unique equipment identifier must be used. In the old days, an
item's part number and serial number were enough. That is no longer the case in the 21st
century. The reasons behind this are quite simple. There could be identical part numbers,
and if the item has different manufacturers, then there could be a duplication of serial
numbers, too. Even if the part numbers are different, when items are sorted by serial number or a similar name there could be duplication of items in the sorted list. It is critical
that each piece of test equipment have a unique identifier. Not only is this a requirement
in most standards and regulations, but it is a fast and easy way to keep your records and
documentation straight, within CAMS and in your filing system for calibration records,
both in hard copy and electronically.

By using a unique identification number for every piece of test equipment that is supported, you eliminate duplication of records. Most CAMS use some type of tag, label, or
ID number that cannot be duplicated in the system. Some organizations either purchase
or manufacture a four-, five-, or six-digit numeral that is affixed to each piece of test equipment. This ID number is then used as the unique identifier within CAMS. It is also used
to identify every record associated with that particular item. This includes calibration and

historical records, vendor and third-party calibration records and certificates, and any other type of record that is generated by that item.

Some areas to consider when looking to purchase ID labels from an outside vendor include:

- Chemical and/or abrasive resistance
- UV resistance (so they will not fade over time and/or with exposure to artificial light or sunlight)
- Heat and cold resistance
- Wet or dry environment usage
- Size of the label
- Bar code or ID number inclusion
- Company or department name inclusion on the label
- Ease of removal if the ID label will often be removed regularly
- Stickiness of adhesive to different surfaces, under different conditions of temperature, humidity, cleaning, sterilization, or the oils from human contact

In addition, some companies use different colored labels to identify where the test equipment can or cannot be used and the type of equipment (used by regular staff, by the calibration department, by Quality Assurance, in a radioactive or restricted environment, and so on). All of these factors should be considered before purchasing ID labels.

Initially, when designing your calibration program, the identification of what items will be tagged and labeled is critical to an efficient, accurate database. The object of this exercise is to identify, label, collect the pertinent information, place the correct label on the item, and complete all of these activities in just one pass through your company when doing the initial inventory.

Having someone familiar with the test equipment used throughout the company can be very beneficial. Test equipment that makes a quantitative measurement is usually calibrated. Any type of test equipment that will be repaired in-house or sent out for external calibration or repair, and all calibration standards should be considered as items that require a tag or label and would be included in the quality calibration system.

An easy way to start your program is to list what needs to be in your master database. Here is an example to help get started:

- The unique identifier or ID number (13169 if it is a five-digit number)
- Model or part number (DU 640, AR1140, P-200)
- Serial number (usually found on the back of or data plate of test equipment, or stamped into the item if too small to have a data plate)
- Type or general category of equipment (spectrophotometer, balance, pipette)
- Calibration interval, in months (1, 3, 6, 9, 12, 18, 24, 36, 48, 60 months)—the longest calibration interval I have seen or heard of is 5 years or 60 months.
- DDC (date due calibration—3/7/2007)
- Preventive maintenance (PM) interval
- PM due date
- Owner (used if different owners of the items exist or may in the future)
- Building number or designation (where the test equipment can be found when calibration or repairs are needed)

- Department number or designation (when required), which is useful when sorting for a particular user or owner
- Room, lab, floor, or area number, which makes finding the test equipment easier
- Manufacturer—be careful here; use the same protocol when listing who made the test equipment. It could have a data plate listing a different manufacturer than is actually printed or stamped on the front case or screen. No matter which way the manufacturer chooses to be listed, use the same way each time you update the database

Other areas that might come in handy in the future for any database might be:

- Name of person assigned to the item, or the user's name
- Bench or rack number where the item is stored or used
- Tool box or equipment shelf where item is stored
- Department name (Operations, R&D, Production, Manufacturing, Quality Assurance, Metrology, Miscellaneous)
- How the test equipment is used (Working Standard, Reference Standard, common test equipment)

Why should you care about this at the beginning of the design of your quality calibration program? By knowing what information will be required for using your new database, it can more easily be collected during the initial inventory, instead of having to redo the inventory to collect more information at a later date. If a company only has 10 items to control and monitor, then this is not a problem. But most companies end up with thousands of pieces of test equipment in their inventory, and having to find and gather information on that many items is easier done once than many times.

Now that a database with headings has been started, a decision must be made on three categories: what items require repair and calibration; what items only will possibly require repairs (these will be designated as Calibration Not Required); and which items will be used as standards (both working and reference—see more on standards in Chapter 7).

All of the items that will require calibration must also be assigned a calibration interval (see Chapter 15 for more information). While doing the initial inventory, if some or all of the items have been previously calibrated, this would be an excellent time to record the date they were calibrated and their next date due calibration (DDC). The CAMS used may or may not have a place to record the calibration date, but if a CAMS has not yet been purchased or the vendor decided upon, it can't hurt to collect that data while performing the inventory.

One of the best determining factors when deciding if test equipment should receive calibration or not can be decided by asking if the item makes a quantitative measurement. Within the biotechnology and pharmaceutical industries there are various pieces of test equipment that fall into both calibrated and not calibrated status. Here is a short list of items that usually require calibration: spectrophotometers, balances, thermal cyclers, water baths, incubators, pH meters, conductivity meters, micro plate readers, power supplies, refrigerators and freezers (at least the temperature probes or system used to monitor their internal readings), dual beam scales, centrifuges, flow meters, micrometers and calipers, and torque testers. There are also items that the average calibration technician does not have the training or standards available to support. But an ID number must still be assigned and the unit put into CAMS. If an outside vendor performs a PM or calibration support, then the information supplied by the vendor needs to be recorded into CAMS.

Some like items that have the same designation in CAMS need to be identified according to their use. A floor-mounted, high-speed centrifuge would normally require calibration and would be designated as such. However, a small bench-top centrifuge used only to quickly mix ingredients that has no readout or adjustable mechanism would be designated as NCR (no calibration required). This type of equipment could be identified in CAMS as a bench top-centrifuge, or table-top centrifuge, while the larger, calibrated units could be designated as high-speed, or floor-mount centrifuges. No matter which way the items are listed, showing a difference in the type of item will save headaches down the road during sorting.

If the items that are chosen to be calibrated have not already received a calibration, during the inventory would be an excellent time to affix a *Do Not Use, Out of Calibration* label.

Also, any items that have been designated as *CBU* or *Standardize Before Use* (see Chapter 10 on labels) should also receive their labels during the initial inventory. Proper planning and a little foresight can save a lot of time and effort in the long run.

While collecting the initial information during this inventory, the ID number and any other labels should also be attached to the test equipment. If the test equipment has been designated as NCR, then affix the proper label, and write the ID number of the test equipment that it is attached to. This will preclude anyone from removing the NCR label and affixing to another unit. Or, if the label falls off, it would be easy to see which item it came from and a new NCR label can be attached.

Once your inventory has been completed and all the required information put into CAMS, or a database, it is time to make some decisions. Sort the database by items that require calibration and items do not. Then, sort the items that require calibration by their type, such as pipette or balance. This will provide an excellent idea of what needs support, and depending on how many calibration technicians are assigned or used, will help determine the number of standards required to support these items. It will help determine which items can be economically calibrated in-house, which will have to either be sent out for support, and which will require a vendor to come to the company to support those items.

It is highly recommended that the inventory be completed before deciding what should be calibrated in-house and what should be sent out for support. Simply asking one of the long-time employees of the company how many pipettes were used is a poor way to make decisions. The first guess I received when starting a calibration program was that there should be no more than 200 pipettes requiring calibration. After the initial inventory was complete, there were found to be more than 1200 pipettes throughout the company. The difference in those numbers made it easy to see that in-house support was not only required but would be a money-saving proposition compared to sending them all out for calibration, adjustment, and preventive maintenance and repair.

Another decision that can be made with more accuracy once your database is complete concerns which and how many standards need to be purchased. The main difference in your standards will be which ones are used daily to calibrate all of the test equipment throughout a company, and which will be used as reference standards to calibrate the working standards (see Chapter 7 for specifics).

As previously mentioned, a quality calibration program must maintain traceability. This can be accomplished only when the standards have traceable calibration back to a national or international standard. This could be accomplished by using standards that receive their calibration from third-party vendors (calibration labs), have traceable calibration from an outside vendor, or are calibrated in-house using reference standards. This can be determined from how many specific pieces of test equipment are supported and the number of standards that are required to support those items. Here is an example with real-life numbers.

A company supports 1,425 items that require some type of temperature calibration. These items range from thermometers (digital and liquid in glass), water baths, autoclaves, centrifuges, thermocouples, thermal cyclers, incubators, heat blocks, pH meters, and thermistors/RTDs used in refrigerators, freezers, and walk-in coolers, to items as simple as temperature chart recorders. They use different types of working standards to calibrate all of those pieces of test equipment. The working standards vary mostly by their accuracy and range, some with specific probes that allow submersion in a water bath for calibration of thermometers, thermocouples, and so on. Others are specific to their purpose such as readings taken at -80 °C. No matter the case, all of the working standards are calibrated in-house using a high-end water bath and temperature standards that are sent out for NIST traceable calibration by a third-party vendor. By ensuring that all in-house calibrations receive at least a TUR of 4:1 or greater, traceability is accomplished, and a significant amount of money is saved year in and year out.

Without the benefit of being able to use the database to see how many items could be supported in-house and compare prices of outside vendor calibration versus the initial cost of purchasing the working standards and reference standard, the cost savings could not be calculated. The return on the investment of the working standards and reference standards was achieved in less than a year.

The same method was used when analyzing standards required for calibration of spectrophotometers, balances, high-speed centrifuges (strobe light), and flow meters. It cannot be stressed enough how important it is to become as self-sufficient in a quality calibration program as possible.

The ability to perform repairs and calibration in-house will save money in time spent sending test equipment out for support, time lost waiting for their return, the actual cost of repairs (labor and in some cases parts), and the cost to the company of not having those items available to their users when needed.

This is all part of organizing your quality calibration program from the start. Design for what will be needed in years to come. Don't just look at the next few months, but plan for expansion in the amount of test equipment to be supported and any ideas that concern upgrading your standards.

A quality calibration program should be included when designing for new systems or programs that will have test equipment. Here are some of the questions that must be asked of the customer when they are looking at the purchase of new test equipment.

Are the required standards needed to support the new items already available within the company? Will the standards that are available provide the necessary traceability requirements? Will the TUR be 4:1 or greater using the available standards? Will additional standards have to be purchased? What will the cost of supporting those standards be? Can they be calibrated in-house or will they need to be sent to an outside vendor? Does the customer really need the accuracy of the test equipment they want or can a less expensive item be purchased to do the same job or function? Are any of the required items located throughout the company but not being used?

To answer this last question, some quality calibration departments maintain a database of unused test equipment that they keep for reissue to their customers. It's possible that certain items are no longer needed when a project or production is finished or canceled; then those items can go back into a reissue stock to help preclude purchasing like items in the future. The list of test equipment available for reissue must either be available to the customer for review, or the calibration department must be on a list of approvers for any requisitions submitted for purchasing of new test equipment. Either way, the information needs to be available to the customer.

12

Calibration Environment

According to RP-6, paragraph 5.11, "The calibration environment need be controlled only to the extent required by the most environmentally sensitive measurement performed in the area."[1] If the equipment being used as a standard, as well as the equipment being calibrated, has an operating range wider than that of the environment where they are used, and a TUR of 4:1 or greater is maintained, there should be no problem meeting any requirements for calibrating under those conditions. However, it is the calibration technician's responsibility to consult the manufacturer's specifications to ensure that those standards can be used in that environment without regard to temperature, humidity, dust, vibration, or other variables.

There are a couple of reasons for this. One is that unless the equipment being calibrated is of such a high tolerance that temperature, humidity, RF, vibration, or dust control might have an impact on its ability to make an accurate measurement, the environment is not a concern. Another reason could be that the vast majority of test equipment is designed and built to be used in a wide variety of environments, but in real life are used in stable facilities where there is no impact on their ability to make accurate measurements.

However, all test equipment has uncertainty and the environment where it is calibrated and/or used can play a critical role in determining the item's known uncertainty. A review of what the standards require concerning environmental controls should remove any doubt as to how critical a part it plays in a company's calibration process.

According to ANSI/ISO/IEC 17025-2005, section 5.3, *Accommodation and Environmental Conditions,* "The laboratory shall ensure that the environmental conditions do not invalidate the results or adversely affect the required quality of any measurement . . . The laboratory shall monitor, control, and record environmental conditions . . . where they influence the quality of the results."[2] ANSI/NCSL Z540-1-1994, section 7, *Accommodation and Environment,* states in part: "Laboratory accommodation (facilities), calibration area, energy sources, lighting, temperature, humidity, and ventilation shall be such as to facilitate proper performance of calibrations/verification. The laboratory shall effectively monitor, control, and record environmental conditions as appropriate." ISO 10012:2003(E), section 6.3.1, reads: "Measuring equipment shall be used in an environment that is controlled or known to the extent necessary to ensure valid measurement results. Measuring equipment used to monitor and record the influencing quantities shall be included in the measurement management system." ANSI/ASQC M1-1996, section 4.4, *Environmental Controls,* states: "Environmental controls shall be established and monitored as necessary to assure that calibrations are performed in an environment suitable for the accuracy required." ANSI/ISO/ASQ Q9001-2000, section 6.4, *Work Environment,* states: "The organization shall determine and manage the work environment needed to achieve conformity to product requirements."[3]

One thing to keep in mind for this handbook is that it is written more for the environment found within a manufacturing or production company than for the environment that is required by a regular calibration laboratory. The main difference is that a calibration laboratory usually is tasked with the calibration of working or reference standards used by other calibration functions, such as those found in a manufacturing or production facility. Because of this, the third-party calibration lab standards are of a higher accuracy (tighter tolerances than those of the test equipment being calibrated against it) and require stricter controls than those used in most manufacturing or production companies.

To give the reader an idea of what this means, the following is a table of the environmental conditions that must be met by calibration laboratories.[4]

Note on laboratory environments:

- Forty-five percent relative humidity is an absolute *maximum* for dimensional areas, to prevent rust and other corrosion.
- Twenty percent relative humidity is an absolute *minimum* for all areas, to prevent equipment damage from electrostatic discharge.
- Temperature stability is the maximum variation over time. This is typically measured at the work surface height.
- Temperature uniformity is the maximum variation through the working volume of the laboratory. This is typically measured at several points over the floor area, between the average work surface height and 1 meter higher.
- The air handling system should be set up so that the air pressure inside the laboratory area is higher than the surrounding area. This will reduce dust because air will flow out through doors and other openings.
- Ideally, a calibration lab should not be on the exterior walls of a building, and it should have no windows. This will make temperature control *much* easier.
- Some measurement areas may have additional limits for vibration, dust particles, or specific ventilation requirements.

Table 12.1 General-purpose calibration laboratories.

Measurement Area	Temperature	Stability and Uniformity	Relative Humidity
Dimensional, optical	20 °C ± 1 °C	± 0.3 °C per hour	20% to 45%
All other areas	23 °C ± 5 °C	± 2.0 °C per hour	20% to 60%

Table 12.2 Standards calibration laboratories, or higher-accuracy requirements.

Measurement area	Temperature	Stability & uniformity	Relative Humidity
Dimensional, Optical	20 °C ± 0.3 °C	± 0.1 °C per hour	20% to 45%
Electrical, Electronic	23 °C ± 1.0 °C	± 1.0 °C per hour	35% to 55%
Physical, Mechanical	23 °C ± 1.5 °C	± 1.5 °C per hour	35% to 55%

- It is important that the working volume of the laboratory is free from excessive drafts. The temperature should be reasonably stable and uniform and any temperature gradients, measured vertically or horizontally, should be small. In order to achieve these conditions, at the standard temperature of 20 °C, good thermal insulation and air conditioning with automatic temperature control is generally necessary.

- The temperature control necessary depends, to some extent, on the items to be calibrated and the uncertainties required. For general gage work the temperature of the working volume should be maintained within 20 °C ± 2 °C. Variations in temperature at any position should not exceed 2 °C per day and 1 °C per hour. These are the minimum expectations for UKAS accreditation.

- For higher grade calibrations demanding smaller uncertainties, such as the calibration of gage blocks by comparison with standards, the temperature of the working volume should be maintained within 20 °C ± 1 °C. Variations in temperature at any position should not exceed 1°C per day and 0.5 °C per hour.

- For the calibration of gage blocks by interferometry, the temperature within the interferometer should be maintained within 20 °C ± 0.5 °C. Variations in temperature shall not exceed 0.1 °C per hour.

- Within the laboratory, storage space should be provided in which items to be calibrated may be allowed to soak so as to attain the controlled temperature. It is most important that, immediately before calibration, time is allowed for further soaking adjacent to, or preferably on, the measuring equipment. Standards, gage blocks, and similar items should be laid flat and side by side on a metal plate for a minimum of 30 minutes before being compared. Large items should be set up and left overnight. This is to ensure that temperature differences between equipment, standards, and the item being measured are as small as possible.

The reader might ask why the more stringent requirements, and how calibration can be performed in a normal production or manufacturing facility without any major impact on accuracy or repeatability. The vast majority of test equipment is manufactured to perform within what might be considered very wide tolerances.

Here are a couple of examples. An analytical balance manufactured by a leading company has the following operating temperature range: 10 °C to 30 °C. Another manufacturer lists the operating temperature range as 5 °C to 40 °C. And still another manufacturer of analytical balances specs theirs with an operating temperature range of 15 °C to 25 °C. As can be seen by these examples, the range in which an analytical balance can be used is not only quite wide, but to be at either extreme would cause the person using the balance to be very cold, or breaking into a sweat. Most of today's modern facilities operate within limits that control the temperature and humidity well within a comfort zone.

That doesn't mean a quality calibration program should not have tolerances listed in the calibration procedures or quality system documentation. To comply with most standards or regulations, limits should be listed when any specific or all general calibration functions should stop. These limits usually equal the tightest tolerances of any standard used by the calibration department. By reviewing the manufacturer's specification table and listing the operating temperature limits for all the department's standards, the lowest and highest tolerance can be found and used. If a certain standard has much tighter temperature tolerances than the other working or reference standards, one could also have specific limits for the use of that standard only.

It must be kept in mind that the vast majority of calibrations take place where the test equipment is used, and so the temperature limits of the working standards must fall within the environment of that particular area. If the temperature and humidity need to be monitored or recorded, there are various chart recorders, data loggers, and environmental monitoring systems available to meet just about anyone's needs or budget.

All of the requirements that were quoted at the beginning of this chapter basically say the same thing. "If the environment where calibration takes place does not influence, impact, degrade, or adversely affect the uncertainty or tolerance of the standard used, or test equipment being calibrated . . . then calibrate away."

NOTES

1. Jay L. Bucher, *The Metrology Handbook* (Milwaukee: ASQ Quality Press, 2004), 103.

2. ANSI/ISO/IEC, *ANSI/ISO/IEC 17025-2005: General requirements for the competence of testing and calibration laboratories* (Milwaukee: ASQ Quality Press, 2005).

3. Bucher, *The Metrology Handbook,* 103.

4. Bucher, *The Metrology Handbook,* 105–106.

13

Calibration Scheduling

Scheduling your workload can be a daunting task, especially for a company that has only a small calibration or metrology department but possibly thousands of items to monitor, repair, and calibrate. An effective system to track, monitor, and forecast the workload is critical to any department's success.

Some of the hurdles that must be overcome include:

- The availability of working or reference standards to accomplish traceable calibrations, while keeping all the calibration technicians gainfully employed
- Unavailability of working or reference standards due to their being sent out for calibration or down for repair and/or calibration
- Scheduling around the customer's workload and production schedules

How does one overcome these hurdles and keep a smooth, efficient operation going with possibly limited resources? The answer is effective calibration scheduling. This can be accomplished only with the use of software (CAMS), flexible calibration technicians, and a good working relationship with the customer. Please keep in mind that during this chapter we are discussing the working environment in a calibration or metrology department operating within a company that only supports their own test equipment.

One of the critical resources that must be in place to have an effective calibration scheduling system is CAMS. Your software must be able to provide an up-to-date forecast of test equipment coming due for calibration. Without this resource, your ability to *see the future* is extremely limited. It will be assumed that your software system not only has the capability to provide a list of the items coming due for calibration, but also can be sorted for various other functions, such as scheduling for each individual calibration technician, showing what is due in particular areas, sections, or buildings, and be flexible enough for each technician to run their own schedules and forecasts.

The smallest group of dates one would like to see for the future should be 30 days. Some organizations like to look at 60 days and 90 days, but for our purposes a 30-day schedule is more than adequate.

Once the schedule has been generated, it should be sorted by the date due calibration. This will quickly give both the calibration technician and the supervisor an idea of critical items that are coming due in the very near future. By generating the list on a more frequent basis (daily or weekly, depending on the size and flexibility of the organization), surprises can be kept to a minimum, and quality customer service can remain at the forefront.

Now that you know what is coming due for calibration, an effective plan for who, what, when, where, and how can be organized. Who will be accomplishing the calibrations can be sorted by areas of responsibilities or types of equipment. What will be calibrated during that particular time frame should be easy enough to analyze. When working and reference standards will be available and the sharing of resources can be sorted out ahead of time. Where the working or reference standards will be, as well as the individual calibration technicians and their calibration procedures, is available for all to see. And finally, how everything comes together to maximize the resources, manpower, and last-minute changes that always happen during the course of each business day.

None of this can be accomplished without the 30-day schedule. Sorting the test equipment into like items (based on type of equipment) is another way to maximize your resources and manpower. Having one calibration technician calibrate all items of a certain type can greatly increase productivity, while immersing them in that type of equipment for familiarity on a large scale. This does not mean that this is all they will do for their entire career. In fact, the reality is likely to be quite the opposite. Possibly next week or next month they could be calibrating an entirely different type of test equipment, while a coworker is doing what they did last week or last month. Cross-training of calibration technicians has been going on for generations and is an efficient use of your resources. However, in some circumstances rare types of test equipment only come due for calibration on an infrequent basis and there might only be one or two of those items within the company. To have competent and trained calibration technicians for these types of test equipment takes long-term planning. However, this is another reason why everyone should be using validated calibration procedures during each and every calibration. No one has to remember anything, except where the calibration procedure is located.

In some instances, 30 days lead time in scheduling your workload will not be enough time for either working or reference standards, or your customer's unique items that cannot be supported in-house that must be sent to a third-party vendor for calibration. Sometimes shipping and time at the vendor might exceed the time you have available. This is when rotatable spares or innovative scheduling come into play.

Rotatable spares refers to a system of having an extra set of items on hand for critical test equipment that, when not working or out for calibration, would keep a system down or product from being manufactured.

An example would be a set of pressure transducers that are critical to a manufacturing system. The system cannot manufacture a product without the transducers in place, but the transducers also provide traceable measurement for the product. The system cannot operate properly without the transducers, but the transducers must be calibrated on a regular basis. This situation is kind of a catch-22.

The solution is simple . . . rotatable spares. A second set of identical transducers is procured. They remain uncalibrated until approximately two months before the first set is due calibration. At that time, set two is sent out for calibration. The two-month time frame is determined by the normal time for shipping and calibration for a set of transducers. Once set two has returned, they are swapped for set one. At that time, set one is checked for operability to determine if an Alert or Action procedure needs to take place. They are supposed to sit on the shelf for 10 months and then be sent out for their calibration. And then the process is repeated. This way, calibration is accomplished only once a year while still maintaining a set of calibrated transducers in the system.

Innovative scheduling is when one can see that a certain standard, whether working or reference, is coming due for calibration and is known to support a large workload and/or critical items within the company. Does this mean you absolutely must purchase another expensive standard simply because the original must be calibrated? Hopefully not.

A couple of options are available. One might be to have a working agreement with another company that has the same or similar standard. Whenever your standard is out for calibration, everyone agrees that your company can use theirs on an as-needed basis. And whenever their standard is out for calibration, you allow them to use your standard when required. Of course, for this to succeed, there has to be good communication between the calibration departments to preclude both standards coming due at the same time.

Another option is to look far enough ahead in your scheduling to see all the items that will be coming due that require a particular working or reference standard. All of those items are then calibrated using the critical/one-of-a-kind standard, and then the standard is sent out for calibration. One drawback to this idea would be a circumstance where one of the items that requires that particular standard for calibration becomes inoperative. Once it is repaired, it would require the standard to receive a traceable calibration before being placed back into service. Hopefully, there are other similar items that could be used in its place until the standard is returned from being calibrated.

A third option is having in place a documented system where you could extend the calibration due date on critical items on an as-needed basis. A critical item would be a piece of test equipment or part of a system that is required for the production of product that is a show stopper for your company. There possibly could be only one of these items in the company due to cost, availability (it is a custom-made unit for a particular job or process), or trained manpower to operate the item is not available during a certain time frame due to illness or injury.

Here is how this type of system would work. A critical item is identified as coming due for calibration. The working or reference standard that is used to calibrate it has been sent out for repair/calibration and is unavailable to perform the required calibration. A couple of processes must be in place to be able to use an extended calibration due date system:

- The item must be identified and the process completed before the date due calibration on the original calibration label and the date listed in CAMS must be the same. Historical data must be researched to show that previous calibration and repair data confirms the reliability of the test equipment.
- The test equipment cannot have its date due calibration extended more than half of its current calibration cycle (if it currently has a 12-month calibration interval, the extended calibration due date could not be more than 6 months into the future).
- Usually the next calibration due date is only far enough into the future to allow the return of the missing standard, plus a small fudge factor.
- A higher level manager, director, or vice president must be the final approval authority to perform this type of calibration extension.
- Documentation must be in place to show:
  - Why this extension is required
  - The original calibration date
  - The original calibration due date
  - The new calibration due date

- The calibration interval
- When the standard was used to calibrate the item after returning from the vendor
- Signature and date of the approval authority

Once the working or reference standard is returned, checked to ensure operability, and has been used to calibrate the critical item, the piece of test equipment is given its regular calibration interval and the next due date is figured accordingly. The documentation is updated with the new information and filed in the extended calibration due date folder for ease in finding during an audit.

It cannot be stressed enough that this type of system should be implemented only on a case-by-case basis, and only for critical-use items. This type of system could be easily abused by less-than-honorable calibration technicians. It could be a fallback solution for a system that is not built on honesty and integrity. But for those that use it as intended, it is an option to having to purchase expensive standards for one-off applications.

Most calibration technicians, supervisors, and managers know that calibration can become mundane and boring. By using creative solutions in your scheduling practices, part of the repetitiveness can be either minimized or removed all together. Here are some suggestions to help with this occupational hazard:

> Let the calibration technicians schedule their own work. Those that have been with the company for some time know and understand what needs to be done, how fast it needs to done, and the ramifications of not getting it done.

> For those new to the career field, a regimented training program, followed by supervisor inputs and suggestions, can quickly get them up to speed on how to effectively schedule their own work. The scheduling and calibration of like items coupled with switching among different types of test equipment to relieve boredom cannot only make your technicians more productive, but also help keep the good calibration technicians with the company for as long as possible.

Nobody enjoyed having a supervisor look over their shoulders while they were making their way up the food chain. So why should they expect anything different from calibration technicians of the 21st century? See Chapter 19 for more on training.

14

Calibration Software

The importance of an efficient scheduling program cannot be overemphasized. However, without the use of some sort of CAMS, scheduling becomes cumbersome at best. Software can be set up to produce a 30-day schedule at the press of a button. Within seconds, the technician, supervisor, or manager knows everything that is coming due in the next month. Not only can work efforts be determined, but focus can be placed on calibrating like items at the same time, or scheduling calibrations prior to sending standards off site. By seeing the future before it gets here, surprises are minimized and production enhanced.

None of us has a crystal ball, or so we have been told. The truth is anyone can access the past, present, and future if they have a quality Computer Automated Management System (CAMS). By using CAMS, both management and the calibration technician can see what happened in the past through historical data files. These files should tell what happened, when it happened, where it happened, and who was involved. They should include the date any piece of test equipment was calibrated, when it is due for calibration, any comments or remarks made by the calibration technician who performed the calibration, and who signed off on the final documentation and process. Any repair costs, time accounting for repair and calibration, and possibly any additional information on pass or fail, accuracy, TURs, and so on, should be included in the historical data.

The same is true for the present. If CAMS is updated and used as it is supposed to be used, a daily log of what has occurred, whether it is showing time only accounting, the repair of test equipment, or the calibration of the same, it should be done on a daily basis. This precludes going into the workplace to calibrate test equipment that was calibrated yesterday; looking for test equipment that was moved last week; or trying to find standards that were either taken out of service because of an out-of-tolerance condition or sent out for calibration by an outside vendor. All of the above information should have been put into CAMS the day it occurred, and CAMS should be your one source for information. It should be a real-time system, accessible to all calibration personnel, from technicians to supervisors to their manager.

Regarding the future, CAMS is critical for determining forecasts, inventory control, and allocation of personnel assets. At the push of a button, most CAMS can show the next 30-day schedule of all items coming due calibration and/or for preventive maintenance. This is an invaluable tool for allocating your assets, in the form of standards (both working and reference standards), calibration procedures, and calibration technicians.

By sorting your 30-day schedule by building location and/or type of test equipment coming due for calibration, one can make the best use of resources and (as the old adage goes) "work smarter instead of longer." Most calibration technicians that have been around for some time know how to calibrate like items in order to minimize set-up and tear-down time. And once a good calibration technician gets on a roll, they can produce far more like items in a given period of time than they can by doing one or two of a certain type of instrument, then calibrating a completely different type of item, and then going back to the original item.

Another time-saving feature of knowing what is coming due in the near future is the ability to multitask during your calibrations. An example of this is when calibrating thermometers in a temperature bath. Once all the slots are full (multiple calibrations of the same type of thermometer because you know which ones are coming due because they showed up on the 30-day, 60-day, or 90-day schedule), the calibration technician can perform calibrations on another type of test equipment while the bath and thermometers are equilibrating. Or in the case of calibrating possibly short-time calibrations (gages, pipettes, calipers, micrometers, and so on), other items that take longer due to temperature changes or lengthy reading times could be in the process of their calibrations (water baths, thermometers, incubators, and so on) while the calibration technician is completing numerous quick turnaround items.

All of us have heard the term doing more with less. The fact is that this is almost impossible. Improvements in technology, innovative ways to use time, resources, and personnel, and instituting such systems as going paperless and wireless, full electronic data collection, automation, and so on, can only take productivity so far. The best that can be accomplished with the resources made available by upper management must be optimized with creative thinking, maximizing the resources that are available, and allowing the calibration technician to be as independent in scheduling and held responsible for production as much as possible.

Any CAMS should have a minimum set of parameters. Following is a list of what should be available in any system.

A master ID menu to include:

- Unique identification number
- A serial number
- Part or model number
- Type of equipment
- Owner of the test equipment (where appropriate)
- Location (building, room, locker, shelf, and so on)
- Department designation (where appropriate)
- Manufacturer

  —Their address
  —Phone numbers—toll free and fax
  —Contact name for service or parts

- Calibration interval
- Service contact information

- Warranty information
- Preventive maintenance interval or information
- Miscellaneous information or remarks section

A report menu with various automated reports that might include:

- 30/60/90 day schedule
- List of all calibrations and/or repairs accomplished during a specific time frame
- Historical list for one item identified by a unique identification number:
  —Might include all previous calibration data
  —All parts and their costs from previous calibrations/repairs
  —All time accounting for that item
- Reverse traceability list—Insertion of a working or reference standard's unique identification number results in a list of all items that were calibrated using that standard, during a specified time frame
- Master ID list
- Overdue calibration list
- List of items that require cosigning or second set of eyes review
- Monthly report list:
  —Shows all calibrations performed during a specific period
  —Shows all repairs and their costs during a specific period
  —Shows all time accounting during a specific period
  —Shows all work (calibration, preventive maintenance, and repairs) by department or division for cost accounting within the company
  —Production totals by individual, section, department, or division

An area for inputting data such as calibrations performed, time accounting, repair actions, and preventive maintenance work performed to include:

- Unique equipment identifier
- Date the action was performed (could be selectable, or if work has to be accounted for on the date performed, non-selectable)
- Next date due calibration once CAMS is updated, helping the calibration technician to put the correct next due calibration date on the calibration label
- Menu for listing the standards that were used during this particular calibration, to include the standard's next calibration date (a second way to check if they are overdue)
- Place to record who performed any repairs, their costs, and the time needed to perform them, with a separate area for comments or remarks
- Place to record who performed the calibration and/or preventive maintenance, with a place for time accounting and comments or remarks

An area for management or supervision to make changes or input data on new technicians that will be using CAMS to include:

- Adding new calibration technicians, supervisors or managers

—A radial button to select which menus they will have access to a place to input their name, employee number, or whatever designated way is used to identify the person

—A place to input a password

• A place to remove or delete obsolete or discarded test equipment that should be removed from the Master ID list

These are only a few suggestions for what should be available in any software purchased over the counter. If a company has the capability to write their own software in-house, this is also a start for what to look for in the scope for developing a new software package.

There are already hand-held devices that allow calibration data to be uploaded and wireless programs that do the same thing at the touch of a button. It is possible to one day have CAMS loaded partially into any computer-operated device, and at the end of the calibration process, the calibration technician simply enters a code and presses a button for the entire process to update the master CAMS software, with no further input from the calibration technician. Science fiction? It has been in place at various organizations for many years.

A biotech company that calibrates its own pipettes and has an in-house software system had integrated the pipette and software package into one system. When the calibration technician has completed the calibration of a pipette, they simply hit the update key and their automated management system updates the database on that particular pipette, gives credit for the time spent, along with cost of parts as appropriate, and the calibration technician moves on to the next pipette. They do not have to enter any other data into a separate system. The entire system has been validated and all hardware qualified. The entire process is paperless and the cost savings in time alone would have paid for a second calibration technician.

15

Calibration Intervals

The biggest factor for deciding what interval to assign your equipment is how often and where it is used. A general rule of thumb is the more often it is used and the higher the criticality of the measurement, the more often it should be calibrated. If the user is checking the item against a check standard or standardizing against a known quantity before use, that would also influence the calibration interval (lengthening it). It is highly recommended that you stipulate that a calibration becomes void the day it is due calibration, at midnight. This removes any doubt or vacillation on the part of the user. Most intervals are broken out into months: 1 month, 3 months, 6 months, 12 months, 24 months, 36 months, and so on. Most calibration management software allows you to select days, weeks, months, and possibly years.

If an item is repaired, adjusted, or simply checked to ensure it is operating properly, then a calibration is performed on the item, CAMS is updated, and a record of the calibration saved. No matter when this happens . . . one week after the previous calibration, one week before it is due calibration, or any amount of time in between, CAMS is updated and a new DDC is established. This becomes the new next due calibration date, not the previous date even though it might be one week before or after. Once the new date is in the system, that is what the user and the calibration department go by for the item's DDC.

Some organizations put policy in place for determining when an item should be recalibrated if it requires a check after being moved or if the customer simply is uncertain about how the item is performing. An example would go something like this. If an item is less than halfway through its calibration interval (12-month interval, and it is being checked at the five-month point) then the item does not need a full calibration, only a check of what is in question at the time. If everything checks out, using traceable standards, then the item is placed back in service without a full calibration. If the same scenario happens and the item is at the seven-month point, no matter if the unit is still good or bad (it would first require adjustment or repair) it would require a full calibration. Of course, in both cases, everything is documented in the remarks or comments section of the record and CAMS.

There are various programs on the market for making interval determinations, but companies on a tight budget can make assessments with the data from their own calibrations. To do this, they will need to know how many items were calibrated during a specific period of time (for example, the previous 12 months). During that same period of time, they must also know how many items did not pass calibration. Once the pass rate

Table 15.1   Sample acceptable limit range.

Type of Equipment	No. of Cals	Out of Tolerance	Pass Rate	Current Interval
Refrigerator-freezer	143	4	97.20%	36 months
Autoclave	21	1	95.24%	6 months
Balance	195	2	98.97%	12 months
Centrifuge	23	0	100.00%	36 months
Conductivity meter	6	0	100.00%	60 months
Digital thermometer	62	2	96.77%	36 months
Thermometer	45	2	95.56%	36 months
Heat block	29	1	96.55%	18 months
Incubator	22	1	95.45%	12 months
Incubator CO_2	41	0	100.00%	9 months
Incubator shaker	23	0	100.00%	12 months
pH meter	14	0	100.00%	36 months
Pipette	2063	28	98.64%	12 months
Power supply	57	2	96.49%	60 months
Spectrophotometer	12	0	100.00%	24 months
Thermal cycler	40	0	100.00%	18 months
Water bath	35	1	97.14%	24 months
Overall Pass Rate:	**98.12%**			

is determined, an acceptable limit range can be assessed for the various types of equipment. Table 15.1 is an example of how this works.

The pass rate is computed from the number of items that were out of tolerance divided by the total number of items calibrated. As can be seen in Table 15.1, various types of test equipment have different calibration intervals. Most started out at 12 months, but some, at the time the calibration program was implemented, had lower calibration interval rates to start with. Over an extended period of time, these lower intervals may have been increased as the pass rate proved that the type of equipment was reliable.

Increasing the calibration interval rates should never be taken lightly. A significant increase, say from 12 months to 24 months, could result in a significant amount of test equipment not meeting their tolerances at the 24-month time period, while having to recall a significant number of items or product. It is prudent to err on the side of caution when extending calibration intervals. Small increases, with the possibility of checking some items during the calibration cycle (an example would be checking at the half and three-quarter points) to ensure they are still within tolerance. If a calibration function has enough data on hand to make a calibration interval increase with an accepted low risk tolerance, then checking the items part way through their cycle may not be warranted.

Table 15.2 shows what a five-year average of pass rates might look like. It might also give the reader an idea of what to do with specific calibration intervals.

For example, all the types of test equipment with a greater than 98% pass rate should have slowly increased their calibration intervals after two to three years of data was available. Of course, there are always exceptions to any rules. Certain types of test equipment may be more critical to a company's production process or used more frequently, or have

Table 15.2 Five-year average pass rate.

Type	Pass Rate 01	Pass Rate 02	Pass Rate 03	Pass Rate 04	Pass Rate 05
Refrigerators, etc.	96.71%	99.45%	96.49%	97.16%	97.20%
Autoclave	97.44%	95.12%	96.00%	96.15%	95.24%
Balance	97.70%	98.39%	98.31%	98.93%	98.97%
Centrifuge	100.00%	100.00%	100.00%	100.00%	100.00%
Conductivity meter	100.00%	100.00%	100.00%	100.00%	100.00%
Digital thermometer	97.87%	97.56%	98.00%	100.00%	96.77%
Thermometer	94.23%	95.45%	94.02%	94.29%	95.56%
Heatblock	95.45%	95.83%	92.86%	93.33%	96.55%
Incubator	88.46%	90.89%	90.91%	95.45%	95.45%
Incubator CO_2	88.57%	91.49%	93.33%	96.43%	100.00%
Incubator shaker	95.65%	94.74%	94.12%	96.15%	100.00%
pH meter	96.77%	100.00%	100.00%	100.00%	100.00%
Pipette	92.89%	94.34%	96.63%	97.97%	98.64%
Power supply	98.44%	100.00%	100.00%	100.00%	96.49%
Spectrophotometer	100.00%	100.00%	100.00%	100.00%	100.00%
Thermal cycler	100.00%	98.48%	100.00%	100.00%	100.00%
Water bath	95.24%	90.67%	95.36%	95.83%	97.14%
Overall Pass Rate:	96.20%	96.61%	96.83%	97.75%	98.12%

an adverse impact if found to be out of tolerance. In a biotechnology or pharmaceutical environment, test equipment making weight and temperature measurements might be more critical to the manufacturing processes, so any test equipment that falls into those categories may not receive calibration interval increases or they might receive them at a reduced rate compared to types of test equipment that are not as critical to the production or manufacturing process.

Some may wonder why the change in pass rates if there is basically no change to the equipment or standards used to calibrate their test equipment. Many variables come into play over the years. A change in personnel could make a difference. If a calibration technician operates with less than full integrity, it would impact the overall pass rate. This same technician might not record data that is out of tolerance in order to avoid additional work, or possibly put in the numbers that would assure an in-tolerance unit simply due to laziness or incompetence. Once this person leaves the company and is replaced by a trained, competent calibration technician who performs calibrations with integrity, the pass rate would seem to go down during the subsequent period of time. But eventually, once the items have received proper alignment or adjustment, regular preventive maintenance inspections, and tender loving care from the new calibration technician, the pass rate will increase to where it should be.

Another determining factor in pass rates can be directly reflected in the type of preventive maintenance program that is in effect at your company. Preventive maintenance will lengthen the usable life of any type of test equipment, while reducing down time (refer to Murphy's Law from Chapter 7) because the items always go bad at the most inopportune times. Preventive maintenance also reduces the expense of repairs by only replacing soon-to-be-worn-out parts, compared to those that have worn out and taken a few other parts with them, usually the higher end parts like motherboards, and so on. More can be found on preventive maintenance in Chapter 20.

Also at the bottom of the previous chart, one can see that there are four different codes for the different percentage rates. Those particular rates come from a broad-based system gained by experience. Generally speaking, if a pass rate of 98% or greater can be achieved, the quality calibration system in place is doing its job. It is extremely difficult to attain a 100% pass rate, no matter the quality of the product or best intentions of the calibration personnel. Things simply go bad, wear out, get dropped, or bad standards are used in the ignorance of the technician doing the job. Whatever the reason or cause, a perfect pass rate over the long haul is almost impossible.

If a quality calibration system can maintain a 98% or greater pass rate, generally speaking, those items should be considered for lengthening of their calibration intervals—possibly even doubling them if the items are not too critical to the operation or production of product. Between 95 and 98% is an area that also might be considered for extending their calibration intervals. In this case, one might consider only increasing by one half the current interval or less than double as in the case of 98% or greater.

If a group of test equipment has less than a 95% pass rate, one might consider leaving the calibration interval where it is. Again, there are always mitigating circumstances for every type of test equipment. They could be critical to the operation, so no matter the pass rate the calibration interval will always remain the same. But one must not forget that the more often an item is calibrated, the more costly in the long term it is to support that item. Calibration costs money in:

- The time it takes to calibrate an item
- The cost of purchasing enough standards to get the job done, while not having standards sit around unused
- The cost of keeping the required parts on hand for repair or preventive maintenance of certain types of test equipment (see Chapter 20 on how a proactive preventive maintenance program can be cost effective)
- The downtime during calibration that the test equipment is not available for use by the required department or personnel

This all equates to performing calibration and preventive maintenance just prior to the test instrument going out of tolerance or being unable to operate properly. Historical data and statistical analysis are required to make these types of decisions. Fortunately, most of this has been done previously, and not much is required by the average calibration technician to keep up with what calibration intervals are needed for the different type of test equipment supported at most companies.

Most manufacturers recommend a calibration interval of one year for their items. Generally speaking this is a good place to start, keeping in mind that when changes to calibration intervals are being considered, the decision making process should be accomplished with enough historical data to make a viable decision. For companies just starting up, they should assume that unless there is historical data available on their particular types of test equipment, they will probably be using the same calibration intervals for some time till they can collect enough information to make a good decision.

Why should a company make their own decision on calibration intervals? Why not use what is working at another company? There could be many factors that make a difference to the operability of test equipment located at different companies. Here are just a few:

- The operating environment could be different.
  —Temperature and humidity have an effect on test equipment.
  —Test equipment functions better in a clean environment than a dirty one. Equipment in a clean room may have a longer calibration interval than those in the same building operating in a dirty environment.
- The training of the personnel using the equipment could be a factor.
  —Technicians and scientist that have been trained on the same test equipment might handle, store, and use the items more efficiently.
  —Untrained or unskilled workers could be harder on test equipment they are unfamiliar with, needing a shorter calibration interval to ensure operability.
- The test equipment could be used for different purposes and not for what it was initially manufactured.
- In some cases, the more an item is used, the more reliable it becomes; with the opposite possibly being true. Less-used test equipment has a higher failure rate due to collection of dust (gets into the micro-circuits), wearing out of parts (seals and o-rings crack and deteriorate due to lack of lubrication), and there have been cases where items sitting on a shelf for extended periods of time become food for various animals, making them unusable when needed.

Should each company start out with one-year calibration intervals for all their test equipment? That is a question for the ages. However, an experienced metrologist or calibration technician can greatly assist in starting a program with reliable calibration intervals based on experience, historical data, and just plain common sense.

Where will calibration intervals be in 10 or 20 years? There is a good chance that they will still be around. Everything wears out and needs to be calibrated at a specific interval. Possibly with the improvement and inclusion of computer software into almost all types of test equipment, a small chip could collect trends, accuracy, and operability and make a determination on the calibration interval for that particular piece of test equipment with the outside influences of environment, handling, and storage built right into the equation.

Part IV
Managing a Quality Calibration Program

16

Continuous Process Improvement

There are two items that must continually be updated: your process improvements and training. The first is discussed in this chapter, and the latter in Chapter 18. If a word or phrase were to be assigned to each chapter in this book, the word for this chapter would have to be ***attitude*** . . . bold and italicized! To be successful, both within your department and when dealing with your customers, there is nothing more important than the group's attitude.

To help ensure that all of the operations used in a calibration function occur in a stable manner, there must be a quality calibration system. The effective operation of such a system should result in stable processes and, therefore, in a consistent output from those processes. Once stability and consistency are achieved, then it's possible to initiate improvements. The more stable and reliable the quality calibration system, the easier it will be to make improvements and spread the benefits out to all of the calibration staff. Improvements in calibration procedures, record keeping, and the actual calibration process will help make the calibration function more efficient and cost effective and increase the production totals as improvements are implemented.

We understand and accept the axiom that "times are changing" and so is everything around us. It doesn't matter if it is calibration procedures, calibration records, our standards, our customer's test equipment, the way we do our strategic planning, or tomorrow's weather . . . everything changes over time. If you believe that something is "written in stone," then you are doomed to failure. Nothing is written in stone, except that quality calibration is the foundation for ensuring accurate, repeatable, and traceable measurements.

The changes that have been made in the world of metrology and calibration over the past several years are amazing. The improvement in standards, the reduction in costs, and the availability to access updated information and data continues to grow.

Nobody is ignorant enough to believe that the quality calibration system alluded to in this book is perfect. Quite the opposite is true. Many individuals of higher intellect, training, and knowledge have come before and put together great systems and programs for calibration. But they have never written them down for others to emulate, evaluate, or improve upon. This is indeed a travesty. That is one of the driving forces behind this book. To put in writing a system that others can copy and improve upon. And the optimum words here are *improve upon* . . . continuously!

Continuous process improvement is not a fly-by-night axiom or flavor of the month. It is what makes any and all processes worth their weight in gold. Life itself shows that without continuous process improvement, all beings, large and small, could not evolve or possibly continue to exist. By not inventing the wheel or fire, humans might not be around today. Animals in all forms have continued to evolve throughout the eons to be adaptable and survive life's dangers.

Such is the nature of all things. Continuous process improvement does not mean we intentionally build or develop something wrong or ineptly. It simply means that we must accept from the start that anything developed from scratch can always be improved upon.

The same can be said for a quality calibration system. As time goes by and the system is put to use, easier ways to do things are found. More productive ways to accomplish calibrations, or to write procedures, or to go paperless come along that had not been thought of before. Simply because they had not been thought of (the technology or training was not previously available) does not mean anyone should think less of themselves for not thinking of that particular improvement earlier. It is just the nature of the beast to constantly find ways to make things better.

Take fire, for instance. We no longer rub two sticks together to get a flame started. In some instances, a flame is not even used or needed. Heat can come from many types of sources. Cooking is probably done more without flame today than with it. Part of this improvement comes from safety (open flame does have some detractors) and part from convenience. But over the long haul, it is an improvement upon an age-old idea of cooking with heat.

There are microwave ovens, convection ovens, portable camp stoves, and cans of Sterno. There is even a system that has no matches, no flames, no electricity, and no water. It is called Sterno Flameless. This is not an advertisement for or endorsement of this system or its parent company. It is just an example of how far we have come since humans first saw fire from the sky (lightning and the byproducts of fire) and starting rubbing two sticks together.

The same is true for a quality calibration system. What is needed, where we can find resources to help achieve it, and experienced help in putting a system together can easily be found in many places. No longer is quality staff tasked with putting together a calibration program from scratch without even knowing the true meaning of calibration. This book could be the foundation for many organizations to start a true quality calibration program where none previously existed. However, they also must keep in mind that there is always room for improvement—always!

No matter how good your system or process may be there can always be improvement. If humanity was at the end of its quest for improvements, then there would be no reason to have a patent office or copyright and trademark laws. We are always inventing new ways to do things, and then along comes another person to improve upon the original idea.

As mentioned in the second paragraph of this chapter, the attitude of your calibration staff will be critical to how they view a quality calibration system and your processes. Do they understand the critical need for a quality calibration function? Do they believe in what they are doing, or are they just going through the motions to get a paycheck?

Everyone realizes that most of the world population works to put food on the table and a roof over their heads. Not many have the luxury to work for free. But to be able to earn a paycheck while doing something worthwhile can surely make the day go by faster and easier. And calibration is critical enough to be well worth the time and effort put in each day by calibration technicians.

How does one motivate a staff to not only come to work with a positive attitude, but also be willing to look at each of their processes with a mind-set of continuous process improvement? Do they really understand the importance of what they do and how it fits into the big picture of the company they work for? Have they been told the importance of the market they are involved in? Do they understand the impact of poor quality on their products or services?

Motivation can be enhanced once the calibration technician understands the importance of what they do on every aspect of their company's products and customers. Let me repeat an example that can be found previously in this book.

A calibration technician staff comes to work every day and does nothing but calibrate pipettes[1] for a biotechnology company. They also calibrate a few odd items, spectrophotometers, thermal cyclers, balances, and water baths. But the vast amount of their time and effort goes into the calibration of pipettes (see Figure 16.1).

How does one motivate these calibration technicians to come to work every day to perform the mundane task of calibrating pipettes? What will the pipettes be used for? There are many functions that must use pipettes to accomplish their jobs. Hospitals, pharmacies, universities, research institutes, biotechnology companies, and pharmaceutical companies all use pipettes in their daily work. Is any of this important or critical to society? Of course they are. One never knows if the pipette they just calibrated will be used for

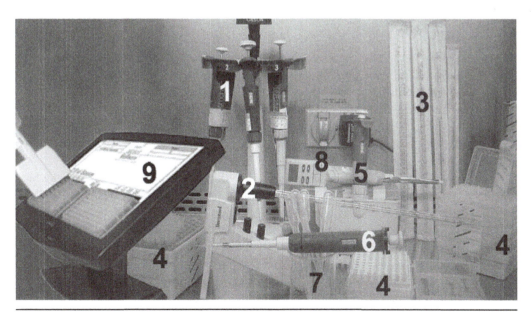

Figure 16.1 Adjustable pipettes.

1. 20-200μl, 20-20μl, 100-1000μl. 2. Graded transfer pipette and electric pipette filler. 3. 25, 10, 5, and 2ml transfer pipettes. 4. Disposable tips for adjustable pipettes. 5. 12-channel adjustable pipette for microplates. 6. Low retention 0.5-10μl adjustable pipette. 7. Squeezable transfer pipettes. 8. Digital adjustable pipette. 9. Light-guided pipetting system.

determining the cause of a patient's illness, dispensing the proper amount of drugs for prescriptions, completing the experiments that allow students to understand the cause and effect of proteins and enzymes, or finding the cure for cancer.

If any one of those situations doesn't get one excited about calibration, then it would be very difficult to motivate a calibration staff to come to work every day. They don't have to work in one of those fields. But somebody has to calibrate all those pipettes.

What about the calibration technicians that are responsible for all the gages and instruments used at an airplane assembly plant? Or an automobile manufacturing facility? Their work is critical for the production of quality parts that affect everyone that flies the skies over us or drives the highways that crisscross America and the world. Hopefully, the wings are not falling off the airliners and the wheels are staying on the vehicles, all due to the diligence and accuracy of the calibration technicians doing their daily work.

Are they doing the same thing that was done 10 or 20 years ago? There had to have been improvements in their processes and systems in order to upgrade to better, faster, and improved products. Somebody had a hand in making those improvements. Somebody saw a way to make it better, faster, and less costly and submitted an improvement to the system to get it implemented. It all falls under the heading of process improvement.

They probably did not get bonuses or anything extra for their time and effort (actually, some might have . . . the glass is always half full). But the attitude to make the process better came from those working the system and seeing where an improvement could be made. The CEO generally doesn't come up with a better way to produce on an assembly line. It is the worker on the line that sees a better, faster way to get the job done. It is the calibration technician that sees how a procedure can be improved and submits the paperwork to get it done.

Attitude. That says it all. Attitude in motivating your staff and yourself. Attitude in opening your mind to see where improvements can be accomplished and then seeing that they get implemented. Part of the process can be found during audits, when an outside set of eyes may see where improvements can be made. But those happen only once or twice a year. The calibration technician working the same old system day in and day out, when properly motivated, should always be on the lookout for an easier, less costly, more productive way to get the job done.

Of course they still have to use calibration procedures (maybe they could be on a laptop or HUD instead of hard copy); and complete the calibration record (going electronic/paperless along with a wireless data collection system instead of hard copy and killing all those trees); and fill out the appropriate calibration labels and CAMS (these could be automated along with the paperless calibration record and time sensitive labels that change color after a predetermined time has passed). The opportunities for making improvements will always be endless. Thus, the chance to make improvements should always be close to the surface when doing your daily work.

Making continuous process improvement a part of your quality system as matter of course can go a long way in helping to motivate the calibration staff. If they observe management and supervision looking for ways to make improvements, and walking the walk, instead of simply mouthing the words, they will know that they have the backing of higher management when it comes time for them to submit their ideas. Just because a certain idea is not used in its entirety doesn't mean that a part of it cannot be used along with somebody else's input. By working together to make the calibration process easier, faster, and more productive, while meeting the requirements of your quality calibration system, there will always be ways to improve how they do their jobs.

NOTE

1. "Pipette," 14 August 2006. http://en.wikipedia.org/wiki/Pipette (22 August 2006). A pipette (also called a pipet) is a laboratory instrument used to transport a measured volume of liquid. Pipettes are commonly used in chemistry and molecular biology research as well as medical tests. Pipettes come in several designs for various purposes with differing levels of accuracy and precision, from single-piece flexible plastic transfer pipettes to more complex adjustable pipettes. A pipette works by creating a vacuum above the liquid-holding chamber and selectively releasing this vacuum to draw and dispense liquid. Pipettes that dispense between 1 and 1000 μl are termed *micropipettes,* while *macropipettes* dispense a greater volume of liquid.

17

Ethics: The Last Frontier

So far this book has shown how to develop and implement a quality calibration program that would meet the requirements and standards of today's stringent world in industry, government, and the military. From the very basics of a quality calibration system, calibration procedures, records, Alert/Action procedures, traceability, uncertainty, the labels, CAMS, the environment, and efficient calibration intervals have all been discussed in detail. Unfortunately, no matter the quality of the calibration program, the good intentions of management or supervision, the unethical behavior and/or conduct of a single calibration technician can place your program, your quality calibration efforts, and even your company's product and/or reputation in jeopardy.

Ethics is one of the five major branches of philosophy that attempts to understand the nature of morality. To distinguish that which is right from that which is wrong. The Western tradition of ethics is sometimes called moral philosophy. Ethics in plain words means studying and analyzing right from wrong; good from bad.[1]

Every day we see reports of the latest findings of illegal and unethical behavior of large and small companies alike. Corruption can be found in sports, politics, our educational systems, financial institutions, and those around us doing their everyday work.

Is this the cost of having to do business in today's fast-paced, hectic world where dog eat dog and the dirtiest trick wins the game? I don't believe that to be the case. All of us have been raised to know the difference between right and wrong. If we haven't, then this book will fall on deaf ears and not be worth the paper it is printed on.

Honesty is the best policy. What goes around comes around. Sound familiar to most of you? We have heard different sayings since we were young children. And why is that? Because they are true! Honesty is still the best policy, in business dealings, in our work environment, and in everything we do as human beings.

But the reader doesn't have to take my word for it. Let's go back in history and see what others have said about honesty, ethics, and doing the right thing.

For Want of a Nail[2]
For want of a nail the shoe was lost.
For want of a shoe the horse was lost.
For want of a horse the rider was lost.
For want of a rider the battle was lost.
For want of a battle the kingdom was lost.
And all for the want of a horseshoe nail.

A clever set of lyrics in "For Want of a Nail," encouraging children to apply logical progression to the consequences of their actions. The rhyme "For Want of a Nail" is often used to gently chastise a child while explaining the possible events that may follow a thoughtless act.

THE HISTORY OF OBLIGATORY ARCHERY PRACTICE!

The references to horses, riders, kingdoms, and battles in "For Want of a Nail" indicate the English origins of the rhyme. One of the English kings did not leave anything to chance! In 1363, to ensure the continued safety of the realm, King Edward III commanded the obligatory practice of archery on Sundays and holidays. The earliest known written version of the rhyme is in John Gower's *Confesio Amantis*[3] dated approximately 1390.

We don't have to go back to the 14th century to see examples of what might happen. Or worse yet, what *did* happen when someone did not do the right thing and compromised their efforts, measurements, or data. It happens every day all around us. All we have to do is look and listen to see the results of these actions.

A pharmaceutical company was caught falsifying data, and when this information came to light they had to recall products (drugs) and spend billions of dollars on lawsuits, court costs, and fines. Numerous companies have been in the headlines during the past few years for corruption, greed, and swindling the public, all for the sake of profit.

A simple measurement that is not taken seriously or made with "only a little error" could result in catastrophe. Here are some examples of what might be:

- The difference of setting the thickness of a sheet metal machine could eventually see the result of not enough strength in the hull of a ship or the skin of an aircraft; the ability to withstand an automobile accident in any type of vehicle, to the strength of a structure to hold up against hurricanes or high winds. The end result? The loss of ships and planes and their valuable crews. The injury and death of men, women, and children when a simple difference of a thousandth of an inch was all it would have taken to prevent any of these possible catastrophes.
- The difference of calibrating an analytical balance in a manufacturing company correctly could mean the production of tainted products that could make the difference between life and death in the manufacturing of drugs, chemicals, and pesticides. A production technician who correctly weighs the product, using an inaccurate balance, could produce either too much of a quantity, or not enough of it, so that the result does not give the expected results to the end user. This could be in a pill or tablet; the amount of chemicals used by any one of us in our daily lives; sprayed on our foods or to eliminate mosquitoes carrying the West Nile virus. The dosage was too weak, so the mosquito was not killed, causing an outbreak of the deadly virus in your neighborhood.

Are there checks and balances in place to keep these from actually happening? I would like to think so. But then, eventually, don't we all end up trusting our coworkers to do the right thing? Is this trust well placed, or are we all preparing for a big disappointment? Or even worse, for the results imagined above?

What does all of this have to do with a quality calibration system? Why is it being included in this book? The average calibration technician has more responsibility than most imagine. Their name or stamp on a calibration label tells the world (or at the very least, the user of that particular piece of test equipment) that the test instrument is not only working properly but accurately. In the vast majority of cases, nobody comes behind the calibration technician to check if what they say is accurate to be really true. Calibration labels are the same as the calibration technicians raising their right hand and swearing that to the best of his or her knowledge, that particular test instrument is accurate, reliable, and producing traceable measurements or readings at the time that they performed that particular calibration. In reality, this is an awesome responsibility. There are few careers where an individual is given this type of responsibility without having a more experienced technician or supervisor sign off on their work. Usually most products have to go through a quality assurance process or testing. Not so with most calibration functions or departments. Their integrity and honesty is a given. Or is it?

What happens when an individual is found to be falsifying data, or, in the old vernacular of PMELs, hot stamping? It could result in their removal from their positions, loss of jobs, and even the end of their careers in calibration. A calibration technician's integrity is the foundation that the career is built on. Once that is questioned or lost, they might as well change careers and move on with their lives; their careers in calibration are over. Nobody has the time or money to double check each and every item calibrated by a "hot-stamper," nor will they be even willing to take the risk of hiring them.

Of course, everyone makes mistakes, does a calculation incorrectly, or interprets their data wrong. When this happens, a responsible calibration technician learns from the mistake, takes responsibility for that action, and carries on. It's the person who is unwilling to accept the great responsibility that has been given them, and does not learn from their mistakes that has no place in the calibration community.

How does a hot-stamper affect the bottom line of the company they are working for? It could affect the company's bottom line in lost sales, lost customers, and in the worst case . . . affect lives. If it seems that this chapter might be getting carried away with how integrity, or the lack of it, can impact a company, then the reader is getting the message loud and clear. It can affect the bottom line of any business, their customers, their employees, and the families of all involved.

There is absolutely no place for dishonest calibration technicians in a quality calibration system. It is as simple as that. No ifs, no buts, no way! Because the opposite end of the spectrum helps demonstrate what an honest calibration technician is responsible for.

Which of the hundreds of pipettes calibrated by the average calibration technician every day will be responsible for helping the scientist discover the cure for cancer? It is as simple as that! Nobody knows. But how can honest, diligent calibration technicians not put their hearts and souls into each and every calibration they perform? It is understood that quantity has to be close to the top of the list of what needs to be done each and every day. But quality comes first, because without quality, all the quantity that is produced in the workplace is irrelevant.

The final result is this: You, the calibration technician, are the front line when it comes to preventing the next great train wreck in the metrology community. Not your supervisor, team leader, manager, or CEO—you! Your honest efforts every day will ensure that the world we live, work, and play in is safe and accurate. Your efforts can and do make a difference.

You should stand a little straighter and taller knowing this. That pipette, dial indicator, micrometer, pressure gage, balance, or insignificant piece of test equipment that was calibrated this morning is the nail that will ensure the kingdom of accurate, traceable, and reliable test equipment is available for all of our customers yet again.

NOTES

1. "Ethics," 22 August 2006. http://en.wikipedia.org/wiki/Ethics (22 August 2006).

2. "For Want of a Nail" Rhyme, 22 August 2006. http://www.rhymes.org.uk/for_want_of_a_nail.htm (22 August 2006).

3. "Confessio Amantis," 2 August 2006. http://en.wikipedia.org/wiki/Confessio_Amantis (22 August 2006). Confessio Amantis ("The Lover's Confession") is a 33,000-line Middle English poem by John Gower, which uses the confession made by an aging lover to the chaplain of Venus as a frame story for a collection of shorter narrative poems. According to its prologue, it was composed at the request of Richard II. It stands with the works of Chaucer, Langland, and the Pearl poet as one of the great works of late 14th-century English literature.

18

Training

As was previously written in Chapter 16, there are two items that must continually be updated: your process improvements and your training. The first was discussed in Chapter 16, and the latter in this chapter.

The following comes straight out of ANSI/ISO/IEC 17025:2005(E), paragraph 5.2:

The laboratory management shall ensure the competence of all who operate specific equipment, perform tests and/or calibrations, evaluate results, and sign test reports and calibration certificates. When using staff who are undergoing training, appropriate supervision shall be provided. Personnel performing specific tasks shall be qualified on the basis of appropriate education, training, experience, and/or demonstrated skills, as required.

The management of the laboratory shall formulate the goals with respect to the education, training, and skills of the laboratory personnel. The laboratory shall have a policy and procedures for identifying training needs and providing training of personnel. The training programme shall be relevant to the present and anticipated tasks of the laboratory. The effectiveness of the training actions taken shall be evaluated.

The laboratory shall use personnel who are employed by, or under contract to, the laboratory. Where contracted and additional technical and key support personnel are used, the laboratory shall ensure that such personnel are supervised and competent and that they work in accordance with the laboratory's management system.

The laboratory shall maintain current job descriptions for managerial, technical, and key support personnel involved in tests and/or calibrations.

The management shall authorize specific personnel to perform particular types of sampling, test, and/or calibration, to issue test reports and calibration certificates, to give opinions and interpretations, and to operate particular types of equipment. The laboratory shall maintain records of the relevant authorization(s), competence, educational and professional qualifications, training, skills, and experience of all technical personnel, including contracted personnel. This information shall be readily available and shall include the date on which authorization and/or competence is confirmed.[1]

That is about as specific as you can get in stating a requirement. It is understood that most calibration departments or functions do not have to meet the requirements of ANSI/ISO/IEC 17025:2005(E); however, they must meet some sort of requirement or standard. Any organization that does not include training as part of their quality calibration system is not going to be meeting any formal standard currently in use today.

Everybody requires training: from their early years when potty training started, to higher education and beyond. Training is part of our lives, both in expanding our knowledge and experience and in learning to adapt to new ideas, concepts, and problems. No one is born knowing how to calibrate test equipment. They are taught through formal education, on-the-job training (OJT), military technical schools, or self-taught through the Internet, home study, or correspondence courses. No matter how one receives new information, training is a lifelong endeavor.

The extent of one's education and training, both formal and hands-on, needs to be documented for several reasons. First, it will preclude having to remember facts that are easily forgotten over time. Second, it is readily available for viewing by inspectors or auditors. Third, it allows anyone to easily see what they are qualified to do, need additional training in, and where they lack knowledge, skill, or experience. A comprehensive training record will ensure the information is available and accurate.

ANSI/NCSL Z540-1-1994 also addresses the requirements for training in section 6, Personnel: "The calibration laboratory shall have sufficient personnel, having the necessary education, training, technical knowledge, and experience for their assigned functions. The calibration laboratory shall ensure that the training of its personnel is kept up-to-date consistent with employee assignments and development. Records on the relevant qualifications, training, skills, and experience of the technical personnel shall be maintained and be available to the laboratory."

ANSI/ISO/ASQ Q10012-2003 continues this theme in section 6.1.2, Competence, and training when it states: "The management of the metrological function shall ensure that personnel involved in the measurement management system have demonstrated their ability to perform their assigned tasks . . . shall ensure that training is provided to address identified needs, records of training activities are maintained, and that the effectiveness of the training is evaluated and recorded."

Ms. Corinne Pinchard said the following in a paper she presented at the 2001 NCSL International Workshop and Symposium in Washington, DC: "How do you save $65,000 a year and get two technicians for the price of one? Easy—you provide in-house training for new technicians on how to calibrate the test equipment used by your company. It is not necessary for your entire core group to be experienced calibration technicians. One experienced person can pass on their knowledge and skills using a well-rounded training program. By providing a solid foundation of knowledge . . . you can have a solid training program in place for minimum cost and effort. A good training program can reap benefits for years to come, especially if it is continually upgraded and improved as circumstances, test equipment, and technician skills change." That was true decades ago, and will continue to be true in the future.

Training and training records consistently fall in the top three areas written up during any audit or inspection. Possibly one of the main reasons is because training is time consuming (some mistakenly believe the company is making no money during training—

nothing could be further from the truth), training is costly (sending personnel to an outside agency to receive qualification or certification), and training records are hard to maintain and nobody really cares (the auditor cares, and it's a requirement in many standards). Generally, trained personnel perform their jobs right the first time (saving the company money by not doing work over), can be used as trainers for untrained personnel (saves the cost of sending everyone out for training), and puts quality at the most important place . . . where the unit is calibrated, not when product is rejected or inspected.

We all know how repetitive and boring it can be to calibrate the same widget 69 times in the same week, while following the same old boring procedure. When someone's life, or the lives of many, or the cure for deadly diseases are on the line, you better be doing it right the first time, and following your procedures is the only way to ensure that is going to happen. And the only way to keep track of your changes and improvements is to document them within your quality system. This is only a small part of the criticality of documenting training.[2]

No matter how old, young, experienced, skilled, or educated a calibration technician may be, they need to continue to train and document that training. This includes managers, supervisors, quality assurance inspectors, and the old guy getting ready to retire some time next year. The production of poor quality work adds expense and time to any process, including calibration. If technicians do not know the correct way to do their jobs, they can only sneak by for only so long. Training removes doubt, instills confidence, and lays the foundation for everyone's skill and experience level.

Can your technicians explain why calibration is needed? Do they understand about the impact that temperature, humidity, magnetism, oil from your hand, dust and dirt, and father time have on any or all test equipment? Just because it turns on does not mean it is working correctly. Many of our customers believe this to be true . . . a simple red light showing the test instrument has power equates to proper, accurate output or readings.

Not only must your calibration technicians be trained in the correct calibration processes and systems, but why it must be accomplished and the benefits, and casualties, of not performing your calibrations correctly.

The documentation of who has been trained, when they were trained, and what they were trained on is not the only documentation required. The actual training program needs to be documented. What training is accomplished, when it should be accomplished, and how it is accomplished should all be part of a well-organized, systematic process of training.

Even if newly hired calibration technicians are experienced and trained when they join your company, there are always areas where they have no experience or education. Different types, brands, or styles of equipment can be confusing and add to the anxiety of a new calibration technician. A comprehensive training program must be used to educate new hires, temporary employees, and those that transfer into your department from some other area of the company, and they all need to be trained from the same training program.

For those who have never started or used a training program, here is a list of items that can be included in a calibration training program. This is not all inclusive and would differ by type of calibrations performed, the end product of your company, or any other differences you may encounter.

A short list might include:

Policies and Practices
Department policy
Hours of operation
Parking
Housekeeping
Time accounting

A list of department policies is just a start of what a newcomer needs to know. It might seem trivial, but having a list of hours of operation, where everyone should park, and general housekeeping rules can take a lot of stress off the new person. Everyone that has worked in a calibration lab, metrology department, or military PMEL assumes everyone has the same experience and knowledge that they do. This is not true. Even when calibration technicians move from one area to another or one job to another, some things are bound to be different. Having them written down saves having to remember them and allows the newcomer to start feeling at home faster than if they have to ask someone every time they have a question. Simply showing the locations of the restrooms, water fountains, vending machines, emergency exits, fire extinguishers, and first-aid kits will start their time in their new positions on a positive note. Having all of this written down for them will also show that management cares and is willing to alleviate as much anxiety as possible.

If the calibration operation is in an environment that uses any type of radioactive materials, formal training needs to be completed for the newcomer. How to perform a wipe test, use of liquid scintillation counters, and/or the proper way to fill out regulator forms should be high on your list of training needs.

Not everyone is familiar with all types of time accounting. Different companies use different systems, so training is mandatory from the start. Some technicians are hired on a wage basis (hourly), while others may be on salary (yearly pay scale). The person earning a wage may not be able to work overtime, whereas the person getting a salary may be expected to work overtime. All of this needs to be explained from the first day, and if the operation is small enough there might not be a human resource person to give the explanation. If this is all written down for the newcomer to read and available when questions arise, stress and anxiety for the new person can possibly be alleviated or at least reduced.

Most calibration technicians that have prior experience have a working knowledge of general housekeeping rules. But different facilities may have specific rules for their areas. It doesn't hurt to have them written down for everyone to follow.

Some calibration functions are not allowed into controlled areas without an escort; have access to some areas at any time, or require special training before being allowed into some areas. All of this information must be written down and everyone trained. As the requirements change, the training manual must also be updated and training of all personnel reaccomplished. Here is a list of items to include information on in your training program:

Quality Documents and Procedure
Calibration systems
Control and use of procedures
CAMS
Control and use of department labels

Control and use of metrology standards
Control and use of metrology department records
Customer service, comments, and suggestions
Procedures when test equipment is found to be out of tolerance
A formal system for updating the calibration training program

One of the positive aspects of a quality calibration system is that nobody is expected to memorize anything. Because everything is written down (or should be), all they have to know is where to find the procedure for accomplishing the task. This may seem a bit redundant, but a procedure that outlines the process for controlling and using your documents is invaluable to the newcomer. Everyone receives training from the same manual or procedure, and nobody has to try to remember what has been told or trained on. When your procedures change, the controlled documents are updated and everyone is trained on the changes, and their training records are annotated to show this has been accomplished. Not only is this a valuable tool within your organization, it is a requirement of most standards and regulations. Here are more ideas to be included in a training program:

Calibration Automated Management System (CAMS)
Master ID menu
Calibration menus
Repair menus
Generating reports
Work orders
Highlighting remarks/comments in CAMS
Short-cycling
Cosigning
Tagging new equipment

Calibration Records
Calibration record location
Calibration records, CAMS, and comments
Saving electronic records
Viewing completed calibration records
Printing calibration records

It doesn't matter if your calibration records are in hard copy or electronic format. Certain aspects are the same. The newcomer needs to know where to find them in order to use them. There should be a written policy on what should or should not go into the comments block of a calibration record. It makes no difference if the record is seldom viewed by anyone outside of the department, or if it is regularly seen be other personnel. Generally, when remarks or comments are made in a calibration record, the same remarks or comments should also be recorded in CAMS. This helps duplicate any and all information that the calibration technician makes. The reason behind this policy is that the supervisor or auditor may not always look at all data during an inspection or review. They may only look at data in CAMS or only the calibration record. No matter which source they use, they should have access to the same information. One must also keep in mind that some systems allow access to all information when using an electronic format. Following is an example.

A well-known biotechnology company is totally electronic (paperless) in their data collection, recording, storage, and archiving. Even when they make comments in only the repair section of their CAMS, it is recorded for posterity. The problem that could arise is that all of the data is available as a read only file for access by anyone in the company. It would not benefit the calibration department to record that they performed a repair and/or calibration while making personal remarks about the users of that test equipment. Because the information is available, it could be embarrassing, even though the information or remarks might be true and factual. Still, common sense and discretion are the better part of valor.

In some quality systems, there is a requirement that any records printed in hard copy must have a statement or stamp applied that says Uncontrolled Document. If this is the case, then stamps must be procured or a watermark established whenever documents are printed or sent electronically to those outside of the calibration function.

Storing and Deleting Equipment
Putting equipment into storage/reissue inventory
Deleting equipment from CAMS
Equipment for reissue

With a calibration department in a company, there could be additional requirements for use and storage of excess test equipment. A written policy should be in place on how to handle this situation. Some items may be too old or abused for reissue or storage. A policy on that should also be written down for all to see.

Some organizations realize that age, use, abuse, and father time can take their toll on test equipment. Having a formal policy in place about when to remove items from inventory and when to store them can only save money for the company. Some programs use a 75% rule. This rule states that when all repair costs, including parts and labor, exceed 75% of the cost of a replacement item, then the old item is scrapped, cannibalized, or discarded. Some other organizations use a 50% rule. No matter which policy is put in place, it should be coordinated ahead of time with management and the calibration technician should be familiar with its specifics.

Homepage or Web page
Homepage within your own company
Web page on the Internet

Some calibration functions maintain their own websites internal to the company (intranets) and some external for the world to view. If the maintenance of either website is the responsibility of calibration technicians, they should receive the training needed to accomplish this function. Whether the training is formal or on-the-job training, it should be recorded and documented the same as any other training.

Other areas that could be in your training manual include: short-cycling protocols, traceability and paper trails, percent of reading as compared to percent full scale, how to attain repeatable readings using analog dials and settings on analog gages, how to determine accuracy and tolerances for the test equipment you support, and possibly what the LSD (least significant digit) is and how it affects your accuracy and readings.

Old-time calibration technicians may know all of the above terms and how to find them. But a lot of untrained old timers are wrong in their knowledge, and when they pass on their incorrect ideas, it only propagates misinformation and inaccuracies in the calibration community. *Short-cycling* refers to shortening the DDC (date due calibration) of a piece of test equipment. There are many reasons why this is beneficial to a company, so only a couple will be reiterated here. An example might be when a group of test equipment is calibrated on the same day. Maybe access to that area is limited, or how the items are used does not allow free and/or easy access on a regular basis. All of the items are calibrated on the same day and CAMS is updated accordingly. However, possibly one of the items is replaced, or repaired and recalibrated. In this case, it would behoove the calibration function to shorten the DDC of that unit to be the same as the original DDC of all the other items in that area. When a calibration report is run to see what is coming due in that area, everything would show up, and the newer item would not fall through the cracks the next time everything comes due for calibration.

Keep in mind short-cycling is not shortening the calibration interval, only shortening the DDC. Another example might be if a large group of items were to be calibrated in a short period of time.

Let's say 250 freezers had temperature-monitoring probes installed for a new environmental monitoring control system. The calibration of these new probes was accomplished over a four week period. The calibration interval for these items was set at 12 months. In order to smooth out the peaks and valleys of the future forecast, the entire group of 250 freezers could have their DDC changed to spread their workload over an entire year, instead of coming due during that particular four-week period. This would be accomplished by evenly spreading the workload out over the following year. All of the items would retain their 12-month calibration interval, but their DDCs would be changed in CAMS to reflect the even spread of DDCs. Each month, approximately 20 of the freezers would come due in CAMS and they would receive a recalibration during that time frame. Once a year has passed, the entire workload would have been spread out, relieving some of the stress of having to do so many during a short period of time. The spreading of the workload could be accomplished by grouping those in the same area where they are located; if responsibility is assigned by calibration technician, they could be grouped by area and technician. No matter how it is accomplished, one would be working smarter instead of longer by using a short-cycle program as needed.

Another use of short-cycling is to smooth peaks and valleys in your forecasted workload. An example can be seen from the forecast chart in Figure 18.1.

One should keep in mind that more weight would be given to items with longer calibration intervals than those with short ones. An example would be items with a calibration interval of six or nine months. These items would probably not be short-cycled because doing so would actually cause them to be calibrated more times in a year than those of longer intervals. The items selected for short-cycling would have their DDCs changed in CAMS to help smooth the peaks and valleys, while also looking at moving like items for ease in calibration. The chart in Figure 18.2 is the same data as Figure 18.1, but the short and tall months have been averaged to see where DDCs can be moved to smooth out the workload. The reader should keep in mind that the chart in Figure 18.1 goes to 600 items, while the chart in Figure 18.2 goes only to 400 after the smoothing has

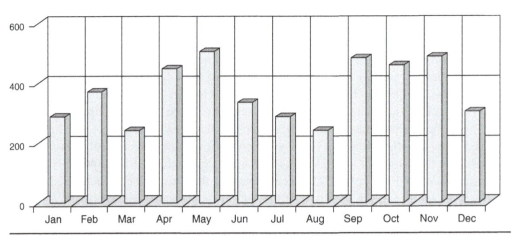

Figure 18.1 Workload forecast.

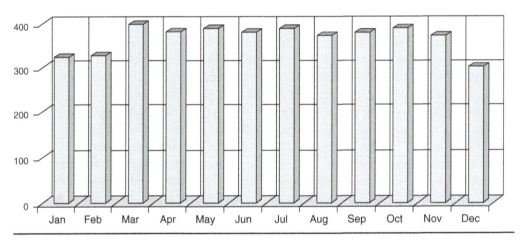

Figure 18.2 Short-cycled workload forecast.

occurred. Some organizations like to keep a lower forecast for December to help in giving extra vacation time during the holidays. Sometimes this cannot be helped, but one can dream.

It is easier when discussing percent of reading compared to percent of full scale when examples are given. An example would be a gage that reads 1000 psi full scale. If the tolerance is percent of reading and the calibration points are at 25%, 50%, 75%, and 95%, then each tolerance would be increasing as the readings are taken from smallest to largest. If we assign a tolerance of ± 10%, then at the 25% point or 250 psi, the tolerance would be ± 25 psi. At 500 psi, the tolerance would be ± 50 psi; at 750 psi the tolerance would be ± 75 psi; and at 950 psi the tolerance would be ± 95 psi.

If we use the same gage and calibration points, but change the tolerance to be ± 10 FS (full scale), then at all the check points, the tolerance would be ± 100 psi. Having a tolerance of ± a certain percentage of full scale increases the overall tolerance. The manufacturer's specifications must be read and interpreted carefully to ensure test equipment receives the correct tolerance, while not increasing the tolerances if interpreted incorrectly. Each user wants to get the most from the test equipment, and the tighter the tolerance, the more you get for your money.

The LSD refers to the least significant digit in any digital readout. In some cases it may change in increments of 1, 2, or 5. This could have a significant affect on the accuracy of the unit being calibrated. It could also influence how a standard is used. If the protocol in the test equipment software rounds according to certain factors, the LSD would move to a higher number or round lower (5 or higher–rounds up, or 4 or lower–rounds down). By increasing a count when using digital numbers, finding out the rounding protocol, or moving to the next significant digit when setting tolerances, one can ensure the highest tolerances while removing doubt about how the user might interpret the data.

This all should be included in your training manual for both initial training and as a refresher over time. Explaining this to your new calibration technicians, and in some cases, to the old timers, could save valuable time down the line when it comes to evaluating if test equipment is in or out of tolerance, or which standards meet your TUR. When TURs get into the 4:1 range, a lot of factors must be considered to meet traceability requirements, and the LSD can play a big factor in meeting or not meeting your TUR.

NOTES

1. *ANSI/ISO/IEC 17025-2005: General requirements for the competence of testing and calibration laboratories* (Milwaukee: ASQ Quality Press, 2005).

2. Jay L. Bucher, *The Metrology Handbook* (Milwaukee: ASQ Quality Press, 2004), 99–100.

19

The Audit

S top! Don't pass GO or collect $200 unless you have read this chapter. It could be the most important chapter in this entire book. I can see *the large question mark* in the bubble above the reader's head. The answer to why is simple. First, read this explanation of a quality system: "The basic premise and foundation of a good quality system is to *say what you do, do what you say, record what you did, check the results, and act on the difference.* Also, for the whole system to work, the organization needs to establish a quality management system to ensure that all operations throughout the metrology department, calibration laboratory, or work area where calibrations are accomplished, occur in a stable manner. The effective operation of such a system will result in stable processes and, therefore, in a consistent output from those processes. Once stability and consistency are achieved, then it's possible to initiate improvements."

How does the organization know that everything is working correctly or according to your written instructions? How do they know where improvements can/should be made? How do they know they're giving their customers the quality service they are paying for? The answer is simple, perform an audit (call it an inspection, review, or check of the quality calibration system).

According to a dictionary definition, an audit is an examination of records or accounts to check their accuracy. It is not vulgar, profane, or have any unwanted calories. But mention the word *audit* to a manager, supervisor, or person in a quality position, and the normal reaction is one of disdain, disgust, and derision. The fact is, nothing could or should be further from the truth. Any organization that is responsive to its customers, both internal and external, and desires to find problems before they affect their products, services, or customers, will conduct audits based on their quality system on a regular basis.

An audit process asks questions, looks at how an organization is supposed to be conducting business to see if it's following its defined procedures, and does this with a mindset of helping itself, its customer base, and its bottom line! An audit is not a bad thing. It costs little in both time and money. Audits afford a quality calibration program the opportunity to correct errors, make improvements, and find areas where it can change for the better before it affects their customers, quality, or certification. It is a self-policing effort.

Are audits a requirement within the various quality systems, regulations, and/or standards? Absolutely. Here are some examples of those requirements. ANSI/ISO 17025-2005(E), paragraph 4.14.1, reads in part: "The laboratory shall periodically . . . conduct internal audits of its activities to verify that its operations continue to comply with the requirements of the quality system and this International Standard. The internal audit programme shall address all elements of the quality system, including the testing and/or calibration activities. It is the responsibility of the quality manager to plan and organize audits as required by the schedule and requested by management. Such audits shall be carried out by trained and qualified personnel who are, wherever resources permit, independent of the activity to be audited."[1] ISO 10012:2003, paragraph 8.2.3, reads: "The metrological function shall plan and conduct audits of the measurement management system to ensure its continuing effective implementation and compliance with the specified requirements. Audit results shall be reported to affected parties within the organization's management. The results of all audits . . . shall be recorded." NCSL International's RP-6, chapter 5.4, reads:

> The calibration control systems should be subject to periodic audits conducted at a frequency and to a degree that will ensure compliance with all elements of the system procedures and documented requirements. It is recommended that a procedure describing the system audits and controls be available and include:
> * Function or group responsible for conducting audits
> * Frequency and extent of audits to ensure compliance with procedures
> * Description of the methods used to ensure that measurements and calibration have been performed with the required accuracy
> * Deficiency reporting and corrective actions required and taken

As a minimum, if there is no internal audit function requirement, a self-inspection program could go a long way in preparing the organization for audits and inspections. By setting up a self-inspection program:

* The organization is showing an effort to find problems
* Sees where it is not meeting the quality system it has in place
* Demonstrates a desire to continuously improve its program through self-initiative
* Finds opportunities before they are found by others
* Makes it proactive instead of reactive to problems and solutions

One option for self-auditing is to follow the *"Say what you do, do what you say, record what you did, check the results, and act on the difference"* theme. Check if the organization is actually following its quality procedures. Do all their records contain the required information, and do they show a paper trail for traceability purposes? If an item was found to be out of tolerance during calibration, was action taken? Was the customer informed, and do they have data to show that it occurred? The more specific the questions, the easier they are to answer. Self-inspections can be an important continuous process improvement, but it takes time, effort, and honesty at all levels. Self-inspections, periodic equipment confidence checks, and so on, all need to be documented if they are to be used in an audit.

Generally, there are three types of audits. An internal audit (first-party audit) is conducted by personnel from within an organization, department, or quality function, and examines its system and records the results for internal eyes only. Internal audits are usually performed by a person who is independent of the department or process that they are auditing to avoid potential conflict of interest. An external audit is conducted by a customer (second-party audit) for the purpose of evaluating its supplier for granting or continuation of business. An external audit is conducted by an auditing agency (third-party audit), with their results being forwarded to the management of the company, department, or organization. Most external audits are performed to see if an organization is in compliance to a specific standard, guideline, or regulation. They can be either subjective or directive in nature. For example, if an organization was audited for compliance to cGMP requirements (FDA), it would be informed of any findings by use of the FDA's Form 483, which is part of the public record. In some cases, accreditation on-site assessments typically require a demonstration of proficiency—over-the-shoulder evaluation/observation. Depending on the inspections criteria, it could include examination of the test equipment, the technician doing the calibration, or the process (calibration procedure, records, documentation, and so on), or all three areas.

How often audits are conducted might depend on who is performing the audit, and for what purpose. An organization may receive an initial audit for ISO 9000:2000 compliance, and then have surveillance audits every six months or yearly. Most internal audits are conducted on a yearly basis unless problems are found, in which case more frequent audits may be performed to ensure improvements are made and conformance is met. Some requirements specify a specific time period for audits, while others leave the time period to the individual company or calibration laboratory.

Once an audit is conducted, the results need to be documented and kept on file for a predetermined amount of time. How long records are maintained should be stated in your records-retention policy. The important point here is that the results are saved for future reference. Follow-up audits to ensure observations, findings, and/or write-ups have been corrected also need to be filed for future needs.

Are any corrective and/or preventive actions identified from the results of the audit? Are any opportunities identified to perform the process in a more efficient manor? Is the proper authority sent the final audit results? Are the people using the quality system aware of the findings and updated on any changes to the system? Is there documentation that supports all of the above? If procedures are changed, are the technicians, supervisors, and manager trained in the new procedures, and their training records updated accordingly? Is there an area in the audit for checking if training records are maintained properly? See Chapter 18 for more information on training and training records.

It is important to assign custodial responsibility for audit discrepancies—the person assigned to follow through with the corrective action plan and the timetable for correcting discrepancies. This should be part of the quality system with timelines, the responsible party that the findings are sent to, and how long the results are maintained.[2]

Did you pass GO and collect your $200? This isn't a game of Monopoly. It is far more serious, and in some instances people's lives, welfare, food, transportation, and much more is at stake. The world of metrology and calibration encompasses more areas than most are aware. To not take the necessary steps to ensure we are doing our jobs to the best of our abilities and not make improvements in a proactive manner is both a waste of time and money.

One of the few ways that we can continually improve our systems and procedures is to review, examine, and/or inspect them on a regular basis. And to do this properly, we should have a second set of eyes do the looking. Bring in an outside party that is neutral to the results, or an individual from another department, section, or from quality assurance if such a department exists at your company that is unfamiliar with what you do or how you do it. Even if you have to pay an outside agency to conduct an audit every couple of years, it is worth the time and effort to ensure your system is operating as it should.

What should be done during an outside audit for compliance to a regulation or a standard? Is there a special set of rules or guidelines? How should everyone act? Is there a protocol or SOP that must be followed? Believe it or not, all these questions have been asked for decades prior to an auditor showing up on the doorstep. The following guidelines can be applied in most situations and modified to work in others.

During most audits, the auditor should be looking at several samples of your quality calibration system. Most audits have to be accomplished during a specific time frame, usually over the course of two or three days, and usually with either one or two auditors. Because time is limited, they will want to review, observe, and/or query about the most important parts of your quality system.

Some of the usual items that they will want to observe are:

- Evidence that the quality calibration system meets the standard or regulation that is stated in your quality manual or that is subscribed to by your organization (ISO 17025-2005, or ISO 9000, or cGMP requirements, as an example). Evidence will be in the form of your documentation and records.
- Evidence that all of your calibration technicians (and anyone else working under your quality system) understand the quality system and how it affects them, their work, and their calibrations.
- Evidence that the quality calibration system follows your written procedures, plans, instructions, and directions. If the quality calibration system states that you must follow written calibration procedures, then the auditor should be able to see each and every calibration technician using calibration SOPs as they go about their daily work. The requirement for calibration records should be self-evident and easy to show the auditor, as well as Alert/Action procedures, forms, and customer notification documentation.
- Evidence that the quality calibration system is effective in providing quality products and services to your customers. High test-equipment pass rates and low or no overdue calibrations are examples of quality that can easily be translated into providing the type of service and support that any customer would accept.

For the auditor to see such evidence, they may observe calibration technicians during routine calibrations or maintenance work. They can also find the evidence in your previous work through your historical calibration records, CAMS data, and by interviewing calibration technicians.

The old adage of working during peacetime the same way we would during war is almost too appropriate for the everyday calibration environment. Your quality calibration system should be set up to work the same as if every day was audit day for your organization. A quality calibration system cannot function efficiently if it only follows its guidelines just before and during an audit.

Follow your procedures, document your calibrations, and perform every function as if an auditor was looking over your shoulder, because in reality they are! Your calibration records will show an auditor how your system really works. And here is why. These are some of the deficiencies commonly found during calibration audits:

- Traceability (the paper trail) is incorrect or nonexistent (if the documentation is not available to show calibration history back to a national or international standard through the working or reference standards used, then legally calibration has not been performed).
- The calibration record's dates do not match CAMS or the calibration label. (They were done after the fact to prepare for an audit.)
- Missing signatures on the calibration record, calibration label, or documentation showing that test equipment was found out of tolerance and notification was provided to the user.
- Inability to perform reverse traceability when a working or reference standard was found to be out of tolerance.
- Overdue calibration stickers on test equipment. (However, a valid calibration record and CAMS historical data could prove that calibration was performed, and that in the haste to turn a lot of production, a calibration technician forgot to put on a new calibration label.)
- Calibration procedures were not followed as a matter course. (The auditor asked the supervisor why a particular calibration technician did not have a calibration procedure available and was told that the person had done hundreds of those calibrations and did not need to have a calibration procedure—major observation. Period.)
- Where required for critical measurements, no records were kept of temperature or humidity (see Chapter 13 for more information).
- Sticky notes in the calibration procedure that differ from the written procedure. (There should be a change control system in place to update all parts of the quality calibration system, and sticky notes/pencil changes are not part of that process.)
- Having empty form fields in calibration records. (Was data *not* recorded, forgotten, or an N/A not put in the field? It is hard to tell because the calibration record was more than two years old and the calibration technician was no longer with the company.)
- Customers are found using test equipment that has overdue calibration labels, or is not even in the quality calibration system. (If records are kept of items that cannot be found when they come due for calibration, or it can be shown that the customer is ordering and receiving test equipment and circumventing the quality calibration system, most auditors will help the calibration function by stressing to upper management that this is a problem that needs to be addressed by management, and out of the hands of the calibration program.)

This is only a very small sampling of observations, write-ups, and nonconformities that have been documented and observed over the years. They are examples of how easy it is to overlook the small things while focusing on the big picture. But quality is everyone's business, not just supervision or management. A quality calibration system starts with the calibration technician, and works up from there. They are the foundation for doing the job correctly each and every day.

An experienced auditor might ask open-ended questions, ones that cannot be answered with a simple yes or no. This will provide them with a better understanding of how well your calibration technicians know and accept your quality program. Some examples of these types of questions might be:

- How do you know how to perform this particular calibration?
- How do you know this function or test meets the quality system or standard?
- Where does it tell you to do this task this way?
- What should you do if an item is found out of tolerance?
- What if the test instrument that is found out of tolerance is your working standard?
- What are the procedures if that happens?

You'll see most auditors taking notes throughout the audit. This is normal and should not be cause for any anxiety. If they need to make an observation or recommendation, they need data to support what happened, when it happened, and what were the circumstances behind the event. Nobody can remember all the details, so they take notes as a matter of course.

In some instances the notes could be general observations or even positive points on how well something works, or how good a process is that they have not seen before. However, please keep in mind that even when an auditor writes down a nonconformity, they are making note of something that does not meet your quality calibration system, not something about the staff or technician. They are auditing your quality system only.

The next logical question would be how to prepare for an audit, either internal or external. Remember the previous statement about working like you are at war even during peacetime? The same applies here more than ever.

- The main responsibility of a calibration technician when preparing for a quality system audit is to know and follow their normal calibration procedures and requirements. To do this they must be familiar with them, and use them every day.
- They must be able to show how to find the correct calibration procedure and calibration records, use CAMS, and locate any other documentation within the quality system that is required for them to perform their jobs.
- Does anyone have personal notes or documentation that they use instead of calibration procedures? Do they have an old procedure squirreled away that could embarrass them or you during an audit? If so, why hasn't this become a part of the formal quality system? There are very good reasons for making it a part:
  —It could be used to help others doing the same job, but who are doing it slower or in a worse way.
  —It could make the difference of the job getting accomplished in the absence of the primary calibration technician.
  —It primarily needs to be done to ensure all calibration procedures are the latest and greatest edition.

Don't be afraid to answer an auditor's questions. Remember, you do not have to memorize anything. That is why everything is written down. You only need to know where to find it (it's in the calibration procedures or the administrative guidelines, on a work instruction or in a continuity book that is easily accessible).

How does one answer an auditor's questions and not swallow their own tongue? Actually, the best answer is usually the easiest. Simply be polite and honest. Generally speaking, an auditor is there to identify problems before they impact your quality or quantity. They are not out to get you or your career. They are usually the same as most of the people they audit, only more experienced with a broader base of knowledge and training.

Of course, we have all seen those that have never calibrated a piece of test equipment in their lives, and never intend to. That's OK, too. If they know how to audit a quality system and its requirements, they can usually do a good job.

Never lie to an auditor. But never tell them more than they are asking for. This is a way for most auditors to find out what part or who is not following the quality system. They let the talkative calibration technician tell them what they cannot readily find during the audit. Remember, only reply to their questions without trying to impress them with your great knowledge or bravado. It is an audit, not a job interview.

In most cases where an auditor is reviewing an entire company and the calibration function is only a small part of the whole, there is usually enough time to only hit a few areas and topics. One of the first questions an auditor will ask is: "May I see your overdue list?" If you have items that are overdue, have they been segregated so that they cannot be used by the customer, or if they are a standard, removed from service to prevent inadvertent usage? Have they been tagged to show they are out of calibration with *Do Not Use* labels?

There are times when there are too many items to remove from service and place in a special location, so the test equipment should have their calibration labels removed and *Do Not Use* labels attached. This should happen as a matter of course, rather than just during audits.

Most auditors will review the calibration function toward the end of their regular audit. The reasoning behind this is that during their inspection of the company, they will be taking a look at the various pieces of test equipment available for all to use. They will write down the identification numbers of some of the test equipment, along with their calibration dates, and their next due dates. Then, when they review the calibration area, they will ask to see the calibration records for those items that they have identified. They will be looking to match the data on the calibration label with the data in the calibration record.

It is curious that a calibration function's CAMS software program is usually not a focus during audits. Part of the reason for this is that the standards and regulations do not absolutely require that CAMS be used. There must be some way to track and monitor test equipment, but it is not spelled out in the standard or any other requirements. Most standards and regulations were written before there were CAMS used by the majority of calibration functions and so they currently do not reference CAMS or automated software.

However, when they ask to see your overdue list, I find it curious to know how one would generate an overdue list unless through a software package. It is hard to believe any organization that supports more than a few pieces of test equipment would be doing so with a hard copy list only. It can be done, but is very time consuming and prone to more human errors than would normally be allowed.

In reality, a person could put the next calibration date into their computer calendar, and then calibrate it when it pops up next year. But that would make proactive work in scheduling the workload extremely difficult. Most companies have enough money in their budget to allow the purchase of some type of CAMS, and that would solve about 99% of the scheduling and forecasting problems.

When the auditor is reviewing your calibration records, this is when they want to see the infamous paper trail, also known as traceability. They will look to see which calibration standards were used, and request to see the latest calibration record on those standards.

The calibration records for your standards should list which standards were used, along with their uncertainty budgets, in order to see if more accurate standards had been used to calibrate your standards. Also, the date due calibration on all of these items must not have been overdue at the time they were used to perform calibrations.

The final result should show that the item in question has traceability all the way back to a national (NIST in the Unites States) or international standard. If all your paperwork is in order, then you should find yourself in the presence of a very impressed auditor.

If your system is paperless and you can readily pull copies of all calibration records for the auditor from any location (this can be accomplished if you are also wireless—see Chapter 5), then they will be truly impressed. Third-party auditors comment how they cannot do that at their company, and they are referring to multi-billion-dollar institutions that have state-of-the-art equipment and innovations. However, their metrology and calibration departments are behind the power curve simply due to the feeling that what they do is not that important. Nothing could be further from the truth. Calibration is the foundation for any company that wants to produce repeatable, reliable, and traceable measurements. Doesn't that seem to be a recurring statement in this book?

Remember that an inspection is an opportunity to see where improvements can be made, a way to find areas that can be improved upon, and a source for making process improvements. If one approaches an audit with the right mind-set, it can be an opportunity for all parties concerned. I wish you the best of luck during your next audit.

NOTES

1. ANSI/ISO/IEC, *ANSI/ISO/IEC 17025-2005: General requirements for the competence of testing and calibration laboratories* (Milwaukee: ASQ Quality Press, 2005), 9.

2. Jay L. Bucher, *The Metrology Handbook* (Milwaukee: ASQ Quality Press, 2004), 89–91.

20

Keeping Your Program Ahead of an Ever-Changing World

The following short subjects are a compilation of systems and programs that have historically shown to improve the bottom line of any company and help any quality calibration program be more efficient, cost less to the organization, and make it as successful as possible.

Customer service. It is written somewhere that "the customer is always right." One could disagree with that statement for several reasons, but the one most appropriate to an imbedded calibration function might be this: "The customer is not always right, but he is always the customer." Bear in mind that there are many ways to approach your customers.

One that comes to mind is using the word "No" to them. Naturally, there will be many times that the customer needs to be told no, but there are different ways to express this word without offending, alienating, or distressing them. If one provides the reason(s) behind using the word no to them, it helps improve understanding and may enhance the relationship. Knowing that the customer's request cannot be met, is a safety problem, or is financially undoable does not give the person giving the answer the power to offend or alienate. Most customers are bright enough about their test equipment to understand why something cannot be accomplished, and appreciate being told the reason behind the answer no.

Another reason to discuss your response is that it might give the customer other options for them to bring to your attention, or give them additional ideas for solutions that the calibration practitioner had not thought to consider.

No matter the reason or reasons for not being able to accomplish the customer's request, there are very few if any reasons not to explain why one cannot do what is asked of them. Nobody likes to be told "No," and have it left at that. The customer is no different in this regard. Tell the reasons, which might even give more options by simply verbalizing them to the customer.

The following example happened decades ago at a U.S. Air Force base overseas. An aircraft test set that was used only on C-130 aircraft came into the PMEL written up for not working properly.

The test set had a multipin plug that was used to connect it to the aircraft and power cart for proper electrical power. While the unit was in PMEL being repaired and/or calibrated, simple pin connectors were used to supply the correct power and pin-outs for checking for the correct voltages. It was common knowledge that it was very dependable and almost never had any problems. This time the write-up stated that it was reading incorrect voltages and not working properly.

After going through the normal calibration procedure and checking to ensure all the fuses and power connections were wired properly, nothing seemed to be amiss. The unit was sent back to the customer with the acronym commonly referred to in the PMEL vernacular as CND (could not duplicate the malfunction).

The next day, the test set was back in PMEL, again written up as not working properly. Obviously something was not working correctly. But the function that was not working could not be duplicated in PMEL. Arrangements were made to walk across the street to the flight line hanger where a C-130 aircraft was parked, and where the test set was to be used. In the presence of the flight line supervisor (who was going to show the lower ranking PMEL technician that he obviously did not know what he was doing), a maintenance specialist attempted to connect the test set to an operating power cart and the aircraft.

When the maintenance specialist connected the power cord, there was a spark and a loud snap. The power was asked to be turned off before any more work was attempted on the test set. Upon close examination, it was found that some very strong individual had attempted to connect the power connect to the test set without properly aligning the pin in the power connector. Because he turned the connector so that the pins were one pin off from the normal connection, none of the functions could work properly within the test set.

Once the connector was properly aligned and connected, and accomplished with the power cart turned off, as per the operating instructions, everything worked properly and they could get the correct readings from the aircraft.

To make a long story short, working with the customer and observing how they followed their procedures (or lack of following procedures), the problem was found, solutions were put in place to keep the problem from reoccurring, and the mission was completed.

If the test set had simply been sent back to them time and time again saying that the problem could not be duplicated, the situation would have continued, and possibly somebody could have been hurt and/or the aircraft damaged. Face to face communication can sometimes remove barriers and solutions found for simply and complex problems.

Become as self-sufficient as possible, as quickly as possible. There are great benefits in becoming self-sufficient in your quality calibration program. What I'm referring to is being able to calibrate the majority of your test equipment in-house instead of sending it to an outside vendor or contracting someone to come to your workplace and calibrate it for you. Being able to repair and calibrate as needed, with traceable standards, will save your company both time and money: Time saved in shipping your test equipment out and back, time lost waiting for another laboratory to complete the work, and the down time your customer suffers during this process. Money is saved in shipping costs, not to mention the high hourly rates for outside calibration and/or vendor service.

Controlling your spare parts and technical/service manuals. If one can organize their spare parts and technical and service manuals into spreadsheet databases, they would have a formal system for being able to find and use them as needed. This allows you to have instant access to location, availability, and inventory control that was previously unknown. Some management software incorporates these functions into their databases, and serious consideration should be given for that option.

Preventive maintenance inspections . . . money in the bank! Just because your equipment is solid state doesn't mean it will last forever. Most of us do not work in a *clean* environment, and neither does our equipment. Over time, air filters are contaminated with dust and dirt, then heat starts to build up inside the equipment, and it's not long before you have a breakdown. We perform preventive maintenance on high-use items at least once a year. How we have put this important program to work for us will be explained.

If this handbook does not have the answer to your calibration or metrology questions, where does one turn for more information? There are numerous organizations available to help answer questions or point in the right direction. A compilation with contact numbers and addresses is furnished for the convenience of the reader, with the understanding that this is not an endorsement of any specific organization or company. The sharing of nonproprietary information throughout the metrology community is one of the hallmarks that help in keep industry, both public and private, on the cutting edge with the latest technology.[1]

American Society for Quality
600 North Plankinton Avenue
Milwaukee, WI 53203
North America: 414-272-8575; Fax: 414-272-1734
International: 800-248-1946

The American Society for Quality (ASQ) has been the world's leading authority on quality since 1946. This professional association creates better workplaces and communities worldwide by advancing learning, quality improvement, and knowledge exchange to improve business results. By making quality a global priority, an organizational imperative, and a personal ethic, ASQ is the community for everyone who seeks technology, concepts, or tools to improve themselves and their world.

Changing the World
A world of improvement is available through ASQ, providing information, contacts, and opportunities to make things better in the workplace, in communities, and in people's lives.

An Impartial Resource
ASQ makes its officers and member experts available to inform and advise the U.S. Congress, government agencies, state legislatures, and other groups and individuals on quality-related topics. ASQ representatives have provided testimony on issues such as training, healthcare quality, education, transportation safety, quality management in the federal government, licensing for quality professionals, and more. Send e-mail to customer service. A Customer Care representative will respond as soon as possible, usually within one business day. Send your message to help@asq.org.

The Measurement Quality Division of ASQ
Members include quality and instrument specialists who develop, apply, and maintain the calibration of measuring equipment and systems, and quality engineers and educators concerned with measurement process capability.

The Measurement Quality Division (MQD) supports, assists, and guides ASQ members and others in the measurement field in the application of both established and innovative tools of measurement and quality. The goal is to improve measurement-based decisions in laboratory, calibration, manufacturing, and management processes at all levels of accuracy.

The MQD supports standards development, disseminates measurement and quality-related information, offers technical support, provides education, sponsors research, fosters professional interaction in the measurement of quality and the quality of measurement, and emphasizes the importance of measurements in the quality process.

National Conference of Standards Laboratories International
2995 Wilderness Place, Suite 107
Boulder, CO 80301-5404
Tel: 303-440-3339; Fax: 303-440-3384

Established in 1961, NCSL International is a professional association for individuals engaged in all spheres of international measurement science. In addition to providing valuable real-time professional career support and advancement opportunities, NCSL International sponsors an annual technical Annual Workshop and Symposium with panels, exhibits, and individual presentations to provide a forum for attendees to exchange information on a wide variety of measurement topics, including implementing national and international standards, achieving laboratory accreditation, new measurement technology, advances in measurement disciplines, laboratory management procedures and skills, equipment management, workforce training, and new instrumentation. NCSL International was formed in 1961 to promote cooperative efforts for solving the common problems faced by measurement laboratories. Today, NCSL International has over 1500 member organizations from academic, scientific, industrial, commercial, and government facilities around the world. This wide representation of experience provides members a rich opportunity to exchange ideas, techniques, and innovations with others engaged in measurement science.

NCSL International is a nonprofit organization whose membership is open to any organization with an interest in the science of measurement and its application in research, development, education, or commerce. Its vision is to promote competitiveness and success of NCSL International members by improving the quality of products and services through excellence in calibration, testing, and metrology education and training. The mission of NCSL International is to advance technical and managerial excellence in the field of metrology, measurement standards, conformity assessment, instrument calibration, as well as test and measurement, through voluntary activities aimed at improving product and service quality, productivity, and the competitiveness of member organizations in the international marketplace.

NCSL International accomplishes its mission through activities whose purposes are to:

- Promote voluntary and cooperative efforts to solve common problems faced by its member organizations.
- Promote and disseminate relevant information that is important to its member organizations. Formulate consensus positions of the membership when requested by outside organizations and government bodies that will serve all or segments of the member organizations.
- Advance the state of the art in metrology and related activities in both the technical and the management area.

- Provide liaison with technical societies, trade associations, educational institutions, and other organizations or activities that have common interests.
- Assess metrology requirements and develop uniform, recommended practices related to the activities of the membership.
- Provide a forum to accomplish the objectives of NCSL through conferences, regional and sectional meetings, committee activities, and publications.
- Serve as an effective channel to assist various national laboratories disseminate information to metrological communities, and to collect and present information to strengthen and improve national measurement systems and the horizontal linkages between these systems.

The Instrumentation, Systems, and Automation Society
67 Alexander Drive
Research Triangle Park, NC 27709 USA
Phone: 919-549-8411; Fax: 919-549-8288
E-Mail: info@ISA.org; www.isa.org

ISA was founded in 1945 as the Instrument Society of America, with a focus on industrial instrumentation. It now has over 39,000 members in 110 countries. ISA is a nonprofit professional society for people in automation and control systems. It promotes innovation and education in application and use of automation and control systems, and in their theory, design, and manufacture. ISA has annual conferences and technical training, and is a publisher of books, magazines, technical standards, and recommended practices.

Institute of Electrical and Electronic Engineers
www.ieee.org
IEEE Instrumentation and Measurement Society (ewh.ieee.org/soc/im)

The American Society for Nondestructive Testing
www.asnt.org/home.htm

The American Society for Nondestructive Testing (ASNT) is the world's largest technical society for nondestructive testing (NDT) professionals. Through its organization and membership, it provides a forum for exchange of NDT technical information; NDT educational materials and programs; and standards and services for the qualification and certification of NDT personnel. ASNT promotes the discipline of NDT as a profession and facilitates NDT research and technology applications. ASNT was founded in 1941 (under the name of The American Industrial Radium and X-Ray Society) and currently boasts an individual membership of nearly 10,000 and a corporate membership of about 400 companies. The society is structured into local sections (or chapters) throughout the world. There are over 75 local sections in the United States and 12 internationally.

The American Society of Test Engineers
www.astetest.org

The ASTE is an all-volunteer, nonprofit corporation with members in 22 states and Canada, including several active chapters. The ASTE is dedicated to the quality, integrity, and advancement of the test engineering profession.

The National Society of Professional Engineers
www.nspe.org

The National Society of Professional Engineers (NSPE) is the only engineering society that represents individual engineering professionals and licensed engineers across all disciplines. Founded in 1934, NSPE strengthens the engineering profession by promoting engineering licensure and ethics, enhancing the engineer image, advocating and protecting PEs' legal rights at the national and state levels, publishing news of the profession, providing continuing education opportunities, and much more.

NOTES

1. Jay L. Bucher, *The Metrology Handbook* (Milwaukee: ASQ Quality Press, 2004), appendix A.

Glossary

Important: Terms that are not in this glossary may be found in one of these primary references:

International Vocabulary of Basic and General Terms in Metrology (called the VIM); BIPM, IEC, IFCC, ISO, IUPAC, IUPAP, and OIML. Geneva: ISO, 1993.

ANSI/NCSL Z540-2-1997, *U.S. Guide to the Expression of Uncertainty in Measurement* (called the GUM). Boulder, CO: NCSL International, 1997.

NCSL Glossary of Metrology-Related Terms, Second edition. Boulder, CO: NCSL International, 1999.

Some terms may be listed in this glossary in order to expand on the definition, but should be considered an *addition to* the references listed above, *not a replacement* of them. (It is assumed that a calibration or metrology activity owns copies of these as part of its basic reference material.)

In technical, scientific, and engineering work, it is important to correctly use words that have a technical meaning. Definitions of these words are in relevant national, international, and industry standards, journals and other publications, as well as publications of relevant technical and professional organizations. Those documents give the intended meaning of the word, so everyone in the business knows what it is. *In technical work, only the technical definitions should be used.*

Many of these definitions are adapted from the references. In some cases, several may be merged to better clarify the meaning or adapt the wording to common metrology usage. *The technical definitions may be different from the definitions published in common grammar dictionaries.* However, the purpose of common dictionaries is to *record the ways that people actually use words*, not to standardize the way the words should be used. *If a word is defined in a technical standard, its definition from a common grammar dictionary should never be used in work where the technical standard can apply.*

GLOSSARY (FROM *THE METROLOGY HANDBOOK*)

accreditation (of a laboratory)—Formal recognition by an accreditation body that a calibration or testing laboratory is able to competently perform the calibrations or tests listed in the accreditation scope document. Accreditation includes evaluation of both the quality management system *and* the competence to perform the measurements listed in the scope.

accreditation body—An organization that conducts laboratory accreditation evaluations in conformance to ISO Guide 58.

accreditation certificate—Document issued by an accreditation body to a laboratory that has met the conditions and criteria for accreditation. The certificate, with the documented measurement parameters and their best uncertainties, serves as proof of accredited status for the time period listed. An accreditation certificate without the documented parameters is incomplete.

accreditation criteria—Set of requirements used by an accrediting body that a laboratory must meet in order to be accredited.

accuracy (of a measurement)—A *qualitative* indication of how closely the result of a measurement agrees with the true value of the parameter being measured. [VIM, 3.5] Because the true value is always unknown, accuracy of a measurement is always an estimate. An accuracy statement by itself has no meaning other than as an indicator of quality. It has quantitative value only when accompanied by information about the uncertainty of the measuring system. *Contrast with:* accuracy (of a measuring instrument)

accuracy (of a measuring instrument)—A *qualitative* indication of the ability of a measuring instrument to give responses close to the true value of the parameter being measured. [VIM, 5.18] Accuracy is a design specification and may be verified during calibration. *Contrast with:* accuracy (of a measurement)

assessment—An examination typically performed on-site of a testing or calibration laboratory to evaluate its conformance to conditions and criteria for accreditation.

best measurement capability—For an accredited laboratory, "the smallest uncertainty of measurement a laboratory can achieve within its scope of accreditation when performing more-or-less routine calibrations of nearly ideal measurement standards intended to define, realize, conserve, or reproduce a unit of that quantity or one or more of its values; or when performing more-or-less routine calibrations of nearly ideal measuring instruments designed for the measurement of that quantity." [EA-4/02] The best measurement capability is based on evaluations of actual measurements using generally accepted methods of evaluating measurement uncertainty.

bias—The known systematic error of a measuring instrument. [VIM, 5.25] The value and direction of the bias is determined by calibration and/or gage R&R studies. Adding a correction, *which is always the negative of the bias*, compensates for the bias. *See also:* correction, systematic error

calibration (1)—(See VIM 6.11 and NCSL pages 4–5 for primary and secondary definitions.) A term that has many different—but similar—definitions. It is the process of verifying the capability and performance of an item of measuring and test equipment by comparison to traceable measurement standards. Calibration is performed with the item being calibrated in its normal operating configuration—as the normal operator would use it. The calibration process uses traceable external stimuli, measurement standards, or artifacts as needed to verify the performance. Calibration provides assurance that the instrument is capable of making measurements to its performance specification when it is correctly used. The result of a calibration is a determination of the performance quality of the instrument with respect to the desired specifications. This may be in the form of a pass/fail decision, determining or assigning one or more values, or the determination of one or more corrections.

The calibration *process* consists of comparing an IM&TE unit with specified tolerances but of unverified accuracy *to* a measurement system or device of specified capability and known uncertainty *in order to* detect, report or minimize by adjustment any deviations from the tolerance limits or any other variation in the accuracy of the instrument being compared. Calibration is performed: *according to* a specified documented calibration procedure, *under* a set of specified and controlled measurement conditions, and *with* a specified and controlled measurement system.

A requirement for calibration does *not* imply that the item being calibrated can or should be adjusted. The calibration process *may* include, if necessary, calculation of correction factors or adjustment of the instrument being compared to reduce the magnitude of the inaccuracy. In some cases, *minor* repair such as replacement of batteries, fuses, or lamps, or minor adjustment such as zero and span, may be included as part of the calibration. Calibration does *not* include any maintenance or repair actions except as noted above. *See also:* performance test, calibration procedure; *Contrast with:* calibration (2) and repair

calibration (2)—Many manufacturers *incorrectly* use the term "calibration" to name the process of alignment or adjustment of an item that is either *newly manufactured* or is *known to be out of tolerance,* or is otherwise in an indeterminate state. Many "calibration" procedures in manufacturers' manuals are actually factory alignment procedures that only need to be performed if a UUC is in an indeterminate state because it is *being manufactured, is known to be out of tolerance,* or *after it is repaired.* When used this way, "calibration" means the same as alignment or adjustment, which are repair activities and excluded from the metrological definition of calibration.

In many cases, IM&TE instruction manuals may use "calibration" to describe tasks normally performed by the operator of a measurement system. Examples include performing a self-test as part of normal operation, or performing a self-calibration (normalizing) a measurement system before use. When "calibration" is used to refer to tasks like this, the intent is that they are part of the normal work done by a trained user of the system. These and similar tasks are excluded from the metrological definition of calibration. *Contrast with:* calibration (1); *see also:* normalization, self-calibration, standardization

calibration activity or provider—A laboratory or facility—including personnel—that perform calibrations in an established location or at customer location(s). It may be external or internal, including subsidiary operations of a larger entity. It may be called a calibration laboratory, shop or department, or a metrology laboratory or department, or an industry-specific name, or any combination or variation of these.

calibration certificate—A calibration certificate is generally a document that states that a specific item was calibrated by an organization. The certificate identifies the item calibrated, the organization presenting the certificate, and the effective date. A calibration certificate should provide other information to allow the user to judge the adequacy and quality of the calibration. In a laboratory database program, a certificate often refers to the permanent record of the final result of a calibration. A laboratory database certificate is a record that cannot be changed; if it is amended later a new certificate is created. *See also:* calibration report

calibration procedure—A controlled document that provides a validated method for evaluating and verifying the essential performance characteristics, specifications, or tolerances for a model of measuring or testing equipment. A calibration procedure documents one method of verifying the actual performance of the item being calibrated against its performance specifications. It provides a list of recommended calibration standards to use for the calibration, a means to record quantitative performance data both before and after adjustments, and information sufficient to determine if the unit being calibrated is operating within the necessary performance specifications. A calibration procedure *always starts with the assumption that the unit under test is in good working order* and only needs to have its performance verified.

Note: A calibration procedure does *not* include any maintenance or repair actions.

calibration program—A process of the quality management system that includes management of the use and control of calibrated inspection, test, and measuring equipment (IM&TE); the process of calibrating IM&TE used to determine conformance to requirements or used in supporting activities. A calibration program may also be called a *measurement management system* (ISO 10012:2003).

calibration report—A document that provides details of the calibration of an item. In addition to the basic items of a calibration certificate, a calibration report includes details of the methods and standards used, the parameters checked, and the actual measurement results and uncertainty. *See also:* calibration certificate

calibration seal—A device, placard, or label that, when removed or tampered with, and by virtue of its design and material, clearly indicates tampering. The purpose of a calibration seal is to ensure the integrity of the calibration. A calibration seal is usually imprinted with a legend similar to "Calibration Void If Broken or Removed" or "Calibration Seal—Do Not Break or Remove." A calibration seal provides a means of deterring the user from tampering with any adjustment point that can affect the calibration of an instrument or detecting an attempt to access controls that can affect the calibration of an instrument. *Note:* A calibration seal may also be referred to as a tamper seal.

calibration standard—An IM&TE item, artifact, standard reference material, or measurement transfer standard *which is designated as being used only to perform calibrations of other IM&TE items.* As calibration standards are used to calibrate other IM&TE items, they are more closely controlled and characterized than the workload items they are used for. Calibration standards generally have lower uncertainty and better resolution than general-purpose items. However, designation as a calibration standard is based on *the use of the specific instrument,* not on any other consideration. For example, in a group of identical instruments one might be designated as a calibration standard while the others are all general purpose IM&TE items. Calibration standards are often called *mea-*

surement standards. (See VIM 6.1 through 6.9, and 6.13, 6.14; and NCSL pages 36–38.) *See also:* standard (measurement)

combined standard uncertainty—The standard uncertainty of the result of a measurement when that result is obtained from the values of a number of other quantities. It is equal to the positive square root of a sum of terms. The terms are the variances or covariances of these other quantities, weighted according to how the measurement result varies with changes in those quantities. [GUM 2.3.4] *See also:* expanded uncertainty

competence—For a laboratory, the demonstrated ability to perform the tests or calibrations within the accreditation scope, and to meet other criteria established by the accreditation body.

For a person, the demonstrated ability to apply knowledge and skills. *Note:* The word *qualification* is sometimes used for this sense, because it is a synonym and has more accepted usage in the United States.

confidence interval—A range of values that is expected to contain the *true* value of the parameter being evaluated with a specified level of confidence. The confidence interval is calculated from sample statistics. Confidence intervals can be calculated for points, lines, slopes, standard deviations, and so on. For an infinite (or very large compared to the sample) population, the confidence interval is

$$CI = \bar{x} \pm t \frac{s}{\sqrt{n}}$$

or

$$CI = p \pm \sqrt{\frac{p(1-p)}{n}}$$

where

CI is the confidence interval,
n is the number of items in the sample,
p is the proportion of items of a given type in the population,
s is the sample standard deviation,
$\bar{x}$ is the sample mean, and
t is the Student's T value for $\alpha/2$ and $(n - 1)$ (α is the level of significance).

correction (of error)—The value that is added to the raw result of a measurement to compensate for known or estimated systematic error or bias. [VIM, 3.15] Any residual amount is treated as random error. The correction value is equal to the negative of the bias.

An example is the value calculated to compensate for the calibration difference of a reference thermometer, or for the calibrated offset voltage of a thermocouple reference junction. *See also:* bias, random error, systematic error

corrective action—Something done to correct a nonconformance when it arises, including actions taken to prevent reoccurrence of the nonconformance. *Compare with:* preventive action

coverage factor—A numerical factor used as a multiplier of the combined standard uncertainty in order to obtain an expanded uncertainty. [GUM 2.3.6] The coverage factor is identified by the symbol k. It is usually given the value 2, which approximately corresponds to a probability of 95%.

deficiency—Nonfulfillment of conditions and/or criteria for accreditation, sometimes referred to as a nonconformance.

departure value—A term used by a few calibration laboratories to refer to bias or systematic error. The exact meaning can usually be determined from examination of the calibration certificate.

equivalence—Acceptance of the competence of other national metrology institutes (NMI), accreditation bodies, and/or accredited organizations in other countries as being essentially equal to the NMI, accreditation body, and/or accredited organizations within the host country. A formal, documented determination that that a specific instrument or type of instrument is suitable for use in place of the one originally listed, for a particular application.

error (of measurement)—In metrology, an *estimate* of the difference between the measured value and the probable true value of the object of the measurement. The error can never be known exactly; it is always an estimate. Error may be systematic and/or random. Systematic error (also known as bias) may be corrected. (See VIM 3.10, 3.12-3.14; and NCSL pages 11–13.) *See also:* bias, correction (of error), random error, systematic error

gage R&R—Gage repeatability and reproducibility study, which (typically) employs numerous instruments, personnel, and measurements over a period of time to capture quantitative observations. The data captured is analyzed statistically to obtain nest measurement capability, which is expressed as an uncertainty with a coverage factor of $k=2$ to approximate 95%. The number of instruments, personnel, measurements, and length of time are established to be statistically valid consistent with the size and level of activity of the organization.

GUM—An acronym commonly used to identify the ISO *Guide to the Expression of Uncertainty in Measurement*. In the United States, the equivalent document is ANSI/NCSL Z540-2-1997, *U.S. Guide to the Expression of Uncertainty in Measurement*.

HIPOT (test)—An acronym for high potential (voltage). A HIPOT test is a deliberate application of extreme high voltage, direct or alternating, to test the insulation system of an electrical product well beyond its normal limits. An accepted guideline for the applied value is double the highest operating voltage plus one kilovolt. Current through the insulation is measured while the voltage is applied. If the current exceeds a specified value a failure is indicated. HIPOT testing is normally done during research and development, factory production and inspection, and sometimes after repair. A synonym is *dielectric withstand testing.*

A high potential tester normally has meters to display the *applied voltage and* the *leakage current at the same time. Caution!* HIPOT testing involves lethal voltages. *Caution!* HIPOT *testing is a potentially destructive test.* If the insulation system being tested fails, the leakage creates a path of permanently lowered resistance. This may damage the equipment and may make it unsafe to use. Routine use of HIPOT testing must be carefully evaluated. *Note:* Hypot® is a registered trademark of Associated Research Corp. and should not be used as a generic term.

IM&TE—Acronym refers to inspection, measuring, and test equipment. This term includes all items that fall under a calibration or measurement management program. IM&TE items are typically used in applications where the measurement results are used to determine conformance to technical or quality requirements before, during, or after a process. Some organizations do not include instruments used solely to check for the presence or absence of a condition (such as voltage, pressure, and so on) where a tolerance is not specified and the indication is not critical to safety. *Note:* Organizations may refer to IM&TE items as MTE (measuring and testing equipment), TMDE (test, measuring, and diagnostic equipment), GPETE (general-purpose electronic test equipment), PME (precision measuring equipment), PMET (precision measuring equipment and tooling), or SPETE (special purpose electronic test equipment).

insulation resistance (test)—A test that provides a qualitative measure of the performance of an insulation system. Resistance is measured in Megohms. The applied voltage can be as low as 10 Volts DC, but 500 or 1000 Volts are more common. Insulation resistance can be a predictor of potential failure, especially when measured regularly and plotted over time on a trend chart. The instrument used for this test may be called an insulation resistance tester or a megohmmeter. An insulation tester displays the insulation resistance in Megohms, and *may* display the applied voltage. *Note:* Megger® is a registered trademark of AVO International and should not be used as a generic term.

interlaboratory comparison—Organization, performance, and evaluation of tests or calibrations on the same or similar items or materials by two or more laboratories in accordance with predetermined conditions.

internal audit—A systematic and documented process for obtaining audit evidence and evaluating it objectively to verify that a laboratory's operations comply with the requirements of its quality system. An internal audit is done by or on behalf of the laboratory itself, so it is a first-party audit.

International Organization for Standardization (ISO)—An international nongovernmental organization chartered by the United Nations in 1947, with headquarters in Geneva, Switzerland. The mission of ISO is *"to promote the development of standardization and related activities in the world with a view to facilitating the international exchange of goods and services, and to developing cooperation in the spheres of intellectual, scientific, technological and economic activity."* The scope of ISO's work covers all fields of business, industry, and commerce except electrical and electronic engineering. The members of ISO are the designated national standards bodies of each country. (The United States is represented by ANSI.) *See also:* ISO

International System of Units (SI)—A defined and coherent system of units adopted and used by international treaties. (The acronym SI is from the French *Systéme International.*) SI is international system of measurement for all physical quantities:

mass, length, amount of substance, time, electric current, thermodynamic temperature, and luminous intensity. SI units are defined and maintained by the International Bureau of Weights and Measures (BIPM) in Paris, France. The SI system is popularly known as the *metric system.*

ISO—A Greek word root meaning *equal.* The International Organization for Standardization chose the word as the short form of the name, so it will be a constant in all languages. In this context, ISO is *not an acronym.* (If an acronym based on the full name were used, it would be different in each language.) The name also symbolizes the mission of the organization, to equalize standards worldwide.

level of confidence—Defines an interval about the measurement result that encompasses a large fraction p of the probability distribution characterized by that result and its combined standard uncertainty, and p is the *coverage probability or level of confidence* of the interval. Effectively, the coverage level expressed as a percent.

management review—The planned, formal, periodic, and scheduled examination of the status and adequacy of the quality management system in relation to its quality policy and objectives, by the organization's top management.

measurement—A set of operations performed for the purpose of determining the value of a quantity. [VIM, 2.1]

measurement system—The set of equipment, conditions, people, methods, and other quantifiable factors that combine to determine the success of a measurement process. The measurement system includes at least the test and measuring instruments and devices, associated materials and accessories, the personnel, the procedures used, and the physical environment.

metrology—The science and practice of measurement [VIM, 2.2].

mobile operations—Operations independent of an established calibration laboratory facility. Mobile operations may include work from an office space, home, vehicle, or the use of a virtual office.

natural (physical) constant—A fundamental value that is accepted by the scientific community as valid. Natural constants are used in the basic theoretical descriptions of the universe. Examples of natural physical constants important in metrology are the speed of light in a vacuum (c), the triple point of water (273.16 K), the quantum charge ratio (h/e), the gravitational constant (G), the ratio of a circle's circumference to its diameter (π), and the base of natural logarithms (e).

NCSL International—Formerly known as the National Conference of Standards Laboratories (NCSL). NCSL was formed in 1961 to *"promote cooperative efforts for solving the common problems faced by measurement laboratories. NCSL has member organizations from academic, scientific, industrial, commercial, and government facilities around the world. NCSL is a nonprofit organization whose membership is open to any organization with an interest in the science of measurement and its application in research, development, education, or commerce. NCSL promotes technical and managerial excel-*

lence in the field of metrology, measurement standards, instrument calibration, and test and measurement." (NCSLI)

nondestructive testing (NDT)—The field of science and technology dealing with the testing of materials without damaging the material or impairing its future usefulness. The purposes of NDT include discovering hidden defects, quantifying quality attributes, or characterizing the properties of the material, part, structure, or system. NDT uses methods such as X-ray and radioisotopes, dye penetrant, magnetic particles, eddy current, ultrasound, and more. NDT specifically applies to physical materials, not biological specimens.

normalization, normalize. *See:* self-calibration

offset—The difference between a nominal value (for an artifact) or a target value (for a process) and the actual measured value. For example, if the thermocouple alloy leads of a reference junction probe are formed into a measurement junction and placed in an ice point cell, and the reference junction itself is also in the ice point, then the theoretical thermoelectric emf measured at the copper wires should be zero. Any value other than zero is an offset created by in homogeneity of the thermocouple wires combined with other uncertainties. *Compare with:* bias

on-site operations—Operations based in or directly supported by an established calibration laboratory facility, but actually performing the calibration actions at customer locations. This includes climate-controlled mobile laboratories.

performance test, performance verification—The activity of verifying the performance of an item of measuring and test equipment, to provide assurance that the instrument is capable of making correct measurements when it is properly used. A performance test is done with the item in its normal operating configuration. A performance test is the same as a calibration (1). *See also:* calibration (1)

policy—Defines and sets out the basic objectives, goals, vision, or general management position on a specific topic. A policy describes what management intends to have done regarding a given portion of business activity. Policy statements relevant to the quality management system are generally stated in the quality manual. Policies can also be in the organization's policy/procedure manual. *See also:* procedure

precision—A property of a measuring system or instrument. Precision is a measure of the repeatability of a measuring system—how much agreement there is within a group of repeated measurements of the same quantity under the same conditions. [NCSL page 26] Precision is *not* the same as *accuracy.* [VIM, 3.5]

preventive action—Something done to prevent the possible future occurrence of a nonconformance, even though such an event has not yet happened. Preventive action helps improve the system. *Contrast with:* corrective action

procedure—Calibration: *see* calibration procedure. General: a procedure describes a specific process for implementing all or a portion of a policy. There may be more than one procedure for a given policy. A procedure has more detail than a policy but less detail than a work instruction. The level of detail needed should correlate with the level of education and training of the people with the usual qualifications to do the work, and the amount of judgment normally allowed to them by management. Some policies may be implemented by fairly detailed procedures, while others may only have a few general guidelines. *See also:* policy

proficiency testing—Determination of laboratory testing performance by means of interlaboratory comparisons.

quality manual—The document that describes the quality management policy of an organization with respect to a specified conformance standard. The quality manual briefly defines the general policies as they apply to the specified conformance standard, and affirms the commitment of the organization's top management to the policy. In addition to its regular use by the organization, auditors use the quality manual when they audit the quality management system. The quality manual is generally provided to customers on request. Therefore, it does not usually contain any detailed policies and never contains any procedures, work instructions, or proprietary information.

random error—The result of a single measurement of a value, minus the mean of a large number of measurements of the same value. [VIM, 3.13] Random error causes scatter in the results of a sequence of readings, and therefore is a measure of dispersion. Random error is usually evaluated by Type A methods, but Type B methods are also used in some situations.

Note: Contrary to popular belief, the GUM specifically does *not* replace *random error* with either Type A or Type B methods of evaluation. [3.2.2, note 2] *See also:* error; *Compare with:* systematic error

repair—The process of returning an unserviceable or nonconforming item to serviceable condition. The instrument is opened, or has covers removed, or is removed from its case and may be disassembled to some degree. Repair includes adjustment or alignment of the item as well as component-level repair. (Some minor adjustment such as zero and span may be included as part of the calibration.) The need for repair may be indicated by the results of a calibration. *For calibratable items, repair is always followed by calibration of the item. Passing the calibration test indicates success of the repair. Contrast with:* calibration (1), repair (minor)

repair (minor)—The process of *quickly* and *economically* returning an unserviceable item to serviceable condition by doing *simple* work using parts *in stock* in the calibration lab. Examples include replacement of batteries, fuses, or lamps, or minor cleaning of switch contacts, repairing a broken wire, or replacing one or two in-stock components. The need for repair may be indicated by the results of a calibration. *For calibratable items, minor repair is always followed by calibration of the item. Passing the calibration test indicates success of the repair. Minor repairs are defined as* repairs that take no longer than a short time as defined by laboratory management; no parts have to be ordered from external suppliers; substantial disassembly of the instrument is *not* required. *Contrast with:* calibration (1); repair

reported value—One or more numerical results of a calibration process, with the associated measurement uncertainty, as recorded on a calibration report or certificate. The specific type and format vary according to the type of measurement being made. In general, most reported values will be in one of these formats:

- *Measurement result and uncertainty:* the reported value is usually the mean of a number of repeat measurements. The uncertainty is usually expanded uncertainty as defined in the GUM.
- *Deviation from the nominal (or reference) value and uncertainty:* the reported value is the difference between the nominal value and the mean of a number of repeat measurements. The uncertainty of the deviation is usually expanded uncertainty as defined in the GUM.
- *Estimated systematic error and uncertainty:* the value may be reported this way when it is known that the instrument is part of a measuring system and the systematic error will be used to calculate a correction that will apply to the measurement system results.

round robin. *See:* interlaboratory comparison

scope of accreditation—For an accredited calibration or testing laboratory, a documented list of calibration or testing fields, parameters, specific measurements, or calibrations and their best measurement uncertainty. The scope document is an attachment to the certificate of accreditation and the certificate is incomplete without it. Only the calibration or testing areas that the laboratory is accredited for are listed in the scope document, and only the listed areas may be offered as *accredited* calibrations or tests. The accreditation body usually defines the format and other details.

self-calibration—A process performed by a user for the purpose of making an IM&TE instrument or system ready for use. The process may be required at intervals such as every power-on sequence, or once per shift, day, or week of continuous operation, or if the ambient temperature changes by a specified amount. Once initiated, the process may be performed totally by the instrument, or may require user intervention and/or use of external calibrated artifacts. The usual purpose is accuracy enhancement by characterization of errors inherent in the measurement system before the item to be measured is connected. Self-calibration *is not* equivalent to periodic calibration (performance verification) because it is not performed using a calibration procedure and does not meet the metrological requirements for calibration. Also, if an instrument requires self-calibration before use, that will also be accomplished at the start of a calibration procedure. Self-calibration may also be called *normalization* or *standardization. Compare with:* calibration (2.B); *contrast with:* calibration (1)

specification—In metrology, a documented statement of the expected performance capabilities of a large group of substantially identical measuring instruments, given in terms of the relevant parameters and including the accuracy or uncertainty. Customers use specifications to determine the suitability of a product for their own applications. A product that performs outside the specification limits when tested (calibrated) is rejected for later adjustment, repair, or scrapping.

standard (document)—(Industry, national, government, or international standard; a *norme*) A *document* that describes the processes and methods that must be performed in order to achieve a specific technical or management objective, or the methods for evaluation of any of these. An example is ANSI/NCSL Z540-1-1994, a national standard that describes the requirements for the quality management system of a calibration organization and the requirements for calibration and management of the measurement standards used by the organization.

standard (measurement)—(Measurement standard, laboratory standard, calibration standard, reference standard: an *étalon*) A *system, instrument, artifact, device*, or *material* that is used as a defined basis for making quantitative measurements. The value and uncertainty of the standard define a limit to the measurements that can be made: a laboratory can never have better precision or accuracy than their standards. Measurement standards are generally used in calibration laboratories. Items with similar uses in a production shop are generally regarded as working-level instruments by the calibration program.

Primary standard: accepted as having the highest metrological qualities and whose value is accepted without reference to other standards of the same quantity. Examples: triple point of water cell; caesium beam frequency standard.

Transfer standard: a device use to transfer the value of a measurement quantity (including the associated uncertainty) from a higher level to a lower level standard.

Secondary standard: the highest accuracy level standards in a particular laboratory, generally used only to calibrate working standards. Also called a reference standard.

Working standard: a standard that is used for routine calibration of IM&TE.

The highest level standards, found in national and international metrology laboratories, are the realizations or representations of SI units. *See also:* calibration standard

standard operating procedure (SOP)—A term used by some organizations to identify policies, procedures, or work instructions.

standard reference material (SRM)—"[I]s a material or artifact that has had one or more of its property values certified by a technically valid procedure, and is accompanied by, or traceable to, a certificate, or other documentation which is issued by NIST. . . . Standard reference materials are . . . manufactured according to strict specifications and certified by NIST for one or more quantities of interest. SRMs represent one of the primary vehicles for disseminating measurement technology to industry." (NIST)

standard uncertainty—The uncertainty of the result of a measurement, expressed as a standard deviation. [GUM 2.3.1]

standardization. *See:* self-calibration

systematic error—The mean of a large number of measurements of the same value, minus the (probable) true value of the measured parameter. [VIM, 3.14] Systematic error causes the average of the readings to be offset from the true value. Systematic error is a measure of magnitude, and may be corrected. Systematic error is also called bias when it applies to a measuring instrument. Systematic error may be evaluated by Type A or

Type B methods, according to the type of data available. *Note:* Contrary to popular belief, the GUM specifically does *not* replace systematic error with either Type A or Type B methods of evaluation. [3.2.3, note] *See also:* bias, error, correction (of error); *Compare with:* random error

test accuracy ratio (TAR)—In a calibration procedure, the ratio of the accuracy tolerance of the unit under calibration to the accuracy tolerance of the calibration standard used. [NCSL, page 2]

$$TAR = \frac{UUT\_tolerance}{STD\_tolerance}$$

The TAR must be calculated using identical parameters and units for the UUC and the calibration standard. If the accuracy tolerances are expressed as decibels, percentages, or another ratio, they must be converted to absolute values of the basic measurement units. In the normal use of IM&TE items, the ratio of the tolerance of the parameter being measured to the accuracy tolerance of the IM&TE. *Note:* TAR may also be referred to as the *accuracy ratio* or (incorrectly) the *uncertainty ratio.*

test uncertainty ratio (TUR)—In a calibration procedure, the ratio of the accuracy tolerance of the unit under calibration to the uncertainty of the calibration standard used. [NCSL, page 2]

$$TUR = \frac{UUT\_tolerance}{STD\_uncert}$$

The TUR must be calculated using identical parameters and units for the UUC and the calibration standard. If the accuracy tolerances are expressed as decibels, percentages, or another ratio, they must be converted to absolute values of the basic measurement units. *Note:* The *uncertainty* of a measurement standard is not necessarily the same as its *accuracy* specification.

tolerance—A design feature that defines limits within which a quality characteristic is supposed to be on *individual parts;* it represents the maximum allowable deviation from a specified value. Tolerances are applied during design and manufacturing. A tolerance is a property of the item being measured. *Compare with:* specification, uncertainty

traceable, traceability—A property of the *result of a measurement,* providing the ability to relate the measurement result to stated references, through an unbroken chain of comparisons each having stated uncertainties. [VIM, 6.10] Traceability is a demonstrated or implied property of *the result of a measurement* to be consistent with an accepted standard within specified limits of uncertainty. [NCSL, pages 42–43] The stated references are normally the base or supplemental SI units *as maintained by a national metrology institute;* fundamental or physical natural constants that are reproducible and have defined values; ratio type comparisons; certified standard reference materials; or industry or other accepted consensus reference standards. Traceability provides the ability to demonstrate the accuracy of a measurement result in terms of the stated reference. Measurement assurance methods applied to a calibration system include demonstration of traceability.

A calibration system operating under a program controls system only implies traceability. Evidence of traceability includes the calibration report (with values and uncertainty) of calibration standards, but *the report alone is not sufficient*. The laboratory must also apply and use the data. *Important:* A calibration laboratory, or a measurement system, or a calibrated IM&TE, or a calibration report or any other thing *is not and cannot be* "traceable to" a national standard. Only the result of a specific measurement can be said to be traceable, provided all of the conditions (discussed previously) are met. Reference to a NIST test number is specifically *not* evidence of traceability. That number is merely a catalog number of the specific service provided by NIST to a customer so it can be identified on a purchase order.

transfer measurement—A type of method that enables making a measurement to a higher level of resolution than normally possible with the available equipment. Common transfer methods are differential measurements and ratio measurements.

transfer standard—A measurement standard used as an intermediate device when comparing two other standards. [VIM, 6.8] Typical applications of transfer standards are to transfer a measurement parameter from one organization to another, or from a primary standard to a secondary standard, or from a secondary standard to a working standard in order to create or maintain measurement traceability. Examples of typical transfer standards are DC Volt sources (standard cells or zener sources), and single-value standard resistors, capacitors, or inductors.

type A evaluation (of uncertainty)—The statistical analysis of actual measurement results to produce uncertainty values. Both random and systematic error may be evaluated by Type A methods. [GUM, 3.3.3 through 3.3.5] Uncertainty can be evaluated only by Type A methods if the laboratory actually collects the data.

type B evaluation (of uncertainty)—Includes any method *except* statistical analysis of actual measurement results. Both random and systematic error may be evaluated by Type B methods. [GUM, 3.3.3 through 3.3.5] Data for evaluation by Type B methods may come from any source believed to be valid.

uncertainty—A property of a measurement result that defines the range of probable values of the measurand. Total uncertainty may consist of components that are evaluated by the statistical probability distribution of experimental data, or from assumed probability distributions based on other data. Uncertainty is an estimate of dispersion; effects that contribute to the dispersion may be random or systematic. [GUM, 2.2.3] Uncertainty is an estimate of the range of values that the true value of the measurement is within, with a specified level of confidence. After an item which has a specified tolerance has been calibrated using an instrument with a known accuracy, the result is a value with a calculated uncertainty. *See also:* type A evaluation; type B evaluation

uncertainty budget—The systematic description of known uncertainties relevant to specific measurements or types of measurements, categorized by type of measurement, range of measurement, and/or other applicable measurement criteria.

UUC, UUT—The unit under calibration or the unit under test—the instrument being calibrated. These are standard generic labels for the IM&TE item that is being calibrated, which are used in the text of the calibration procedure for convenience. (Also may be called DUT [device under test] or EUT [equipment under test].)

validation—Substantiation by examination and provision of objective evidence that verified processes, methods, and/or procedures are fit for their intended use.

verification—Confirmation by examination and provision of objective evidence that specified requirements have been fulfilled.

VIM—An acronym commonly used to identify the ISO *International Vocabulary of Basic and General Terms in Metrology*. (The acronym comes from the French title.)

work instruction—In a quality management system, the detailed steps necessary to carry out a procedure. Work instructions are used only where they are needed to ensure the quality of the product or service. The level of education and training of the people with the usual qualifications to do the work must be considered when writing a work instruction. In a metrology laboratory, a calibration procedure is a type of work instruction.

Index

About the Author

Jay L. Bucher started his calibration and metrology career in 1971 with the U.S. Air Force's Precision Measurement Equipment Laboratories (PMEL) program. He advanced through increasingly more challenging positions, from calibration technician to section supervisor and quality assurance to calibration laboratory manager. In 1994 he was selected to upgrade the capabilities of the Indonesian Air Force's PMEL program. Jay trained its officers and NCOs in all aspects of PMEL management and established its initial quality assurance and scheduling programs.

Jay's accomplishments while serving in the Air Force include:

- April 1972: Branch Student of the Month, Avionics/PMEL, Lowry Technical Training Center, Lowry AFB, CO
- May 1972: Distinguished Graduate from PMEL technical training school, Lowry Technical Training Center, Lowry AFB, CO
- October 1975: Received an Excellent rating during SAC MSET audit of 55 SRW, Offutt AFB, NE (only three were given during that audit)
- December 1977: Distinguished Graduate from 22nd Air Force NCO Leadership School, Norton AFB, CA
- June 1981: NCO of the Quarter, 316th FMS, Yokota AB, Japan
- November 1981: NCO of the Year, 316th FMS, Yokota AB, Japan
- April 1982: Awarded the Air Force Commendation Medal for duty performed at Yokota AB, Japan from September 1976 to March 1982
- July 1982: Avionics Maintenance Technician of the Month, 316th FMS, Yokota AB, Japan
- August 1982: NCO of the Quarter, 316th FMS, Yokota AB, Japan
- August 1982: NCO of the Quarter, 316th TAG, Yokota AB, Japan
- August 1982: NCO of the Quarter, 374th TAW, Clark AB, Republic of the Philippines
- August 1984: Awarded the Air Force Commendation Medal (First oak leaf cluster) for duty performed at Yokota AB, Japan, from April 1982 to June 1984
- February 1985: Graduate of the PACAF NCO Academy, Kadena AB, Okinawa
- August 1985: Awarded the Air Force Commendation Medal (Second oak leaf cluster) for duty performed at Kunsan AB, Republic of Korea, from July 1984 to July 1985
- August 1986: NCO of the Quarter, 432nd CRS, Misawa AB, Japan
- August 1989: Awarded the Meritorious Service Medal for outstanding service performed at Misawa AB, Japan from July 1985 to May 1989
- November 1989: Graduate of the SrNCO Academy correspondence program

- October 1991: SrNCO of the Quarter, 374th MXS, Yokota AB, Japan
- October 1991: SrNCO of the Quarter, 374th TAW/MA, Yokota AB, Japan
- September 1993: Managing Editor/Publisher of *The PMEL Gazette*
- February 1995: Awarded the Air Force Achievement Medal for outstanding achievement as team chief, PMEL Technical Assistance Team while TDY to Iswahyudi AB, Madiun, Indonesia
- August 1995: Awarded the Meritorious Service Medal (First oak leaf cluster) for outstanding service performed at Yokota AB, Japan from June 1989 to August 1995

Retiring from the Air Force after 24 years of service, Jay then spent time working as the senior metrologist for the Royal Saudi Air Defense Force PMEL in Jeddah, Saudi Arabia. He joined Promega Corporation (a biotechnology company) in 1997, where he developed and implemented all facets of an ISO 9001- and cGMP-compliant program for its metrology department. He took the department paperless in 1999 and wireless in 2005. His department was rated "Best-in-Class" for three consecutive years during annual quality system reviews, while supporting more than 6,700 items with a zero-overdue calibration rate for the past nine years with a staff of only four.

In 2000, Jay founded the Madison, Wisconsin, section for the National Conference of Standards Laboratories (NCSL) International and was its section coordinator for five years. He is now the NCSL regional coordinator for the north-central division.

Jay established Bucherview Metrology Services in 2002 and has since consulted with clients ranging from the National Institute of Standards and Technology (NIST) to a third-party one-man calibration function. He has taught the requirements of calibration in a cGMP environment (FDA compliance) for the Madison Area Technical College to include quality systems, traceability of measurements, uncertainty budgets, record maintenance and documentation, and compliance with 21CFR part 11.

Most recently, Jay was a major contributor in the creation of ASQ's Certified Calibration Technician (CCT) Program and a subject matter expert (SME) for its CCT exam. In 2004 he received the Max J. Unis Award by the ASQ Measurement Quality Division (MQD), its highest honor, in recognition of outstanding contributions to the metrological community. In 2005 he was given the NCSL Region/Section Coordinator of the Year Award for the Central Division and selected as a finalist for Test Engineer of the Year by *Test & Measurement World* magazine.

Jay is an ASQ Certified Calibration Technician and an officer with the MQD. He is the managing editor/publisher of the division's quarterly newsletter, *The Standard*, and was editor and co-author of *The Metrology Handbook* (ASQ Quality Press, 2004). He has presented papers at the NCSL Madison section, as well as at NCSL national and international conferences. He has been published in NCSL conference proceedings, as well as Measurement Science Conference, *Cal Lab Magazine, The Standard,* and *Quality Progress*.

Jay lives in De Forest, Wisconsin, with his wife and daughter.